T0330283

Adaptation to Climate Change in Asia

Adaptation to Climate Change in Asia

Edited by

Sushil Vachani

Boston University, USA

Jawed Usmani

Government of Uttar Pradesh, India

Edward Elgar

Cheltenham, UK • Northampton, MA, USA

Published by
Edward Elgar Publishing Limited
The Lypiatts
15 Lansdown Road
Cheltenham
Glos GL50 2JA
UK

Edward Elgar Publishing, Inc.
William Pratt House
9 Dewey Court
Northampton
Massachusetts 01060
USA

A catalogue record for this book
is available from the British Library

Library of Congress Control Number: 2013957785

This book is available electronically in the ElgarOnline.com Business Subject Collection, E-ISBN 978 1 78195 473 7

ISBN 978 1 78195 472 0

Typeset by Servis Filmsetting Ltd, Stockport, Cheshire
Printed and bound in Great Britain by T.J. International Ltd, Padstow

Contents

Contributors

BANGLADESH

Motasim Billah is a Research Associate at the Refugee and Migratory Movements Research Unit (RMMRU), University of Dhaka. He received an MSc from the Institute of Development Studies, University of Sussex, where he focused on climate change and development. His areas of interest are climate change, migration, livelihood and poverty, community-based adaptation to climate change and climate change finance and negotiation. At RMMRU, he has worked on several primary research projects on migration, development and climate change involving extensive fieldwork. He worked as Research Associate in a collaborative research project 'Migrating Out of Poverty', based at the University of Sussex. He co-authored the Dhaka workshop report of Foresight, a project of the UK Government Office for Sciences, *Global Environmental Migration: Bangladesh, Low Elevation Coastal Zones and Islands*. He is co-author of a forthcoming journal article 'Climate change related migration in rural Bangladesh: a behavioural model' to be published in the *Journal of Population and Environment*. Billah has first-hand experience of research-based policy advocacy campaigns and policy analysis. He represented RMMRU in two inter-ministerial committees on determining the minimum wage of migrant workers and the revision of the 1981 immigration ordinance.

Tasneem Siddiqui is Professor of Political Science at the University of Dhaka, where she is also the founding Chair of the Refugee and Migratory Movements Research Unit (RMMRU). Siddiqui's books, monographs and research papers (http://www.rmmru.org) on temporary labour migration of women from Bangladesh, migrant workers' remittance, the Bangladeshi diaspora and development and labour recruitment process have successfully contributed to concrete policy changes such as lifting the ban on female labour migration from Bangladesh, recruitment of workers through computerized databases and liberalization of central bank policies on remittance transfer. Her current research concentrates on the impact of climate change on migration. Through policy advocacy backed by research she aims to convince ministries that by focusing on the environment, disaster risk reduction, labour and expatriate welfare, migration should not be seen as a failure to adapt locally but one of the many adaptation tools for affected people. Siddiqui was one of the key members that drafted the National Overseas Employment Policy of Bangladesh, which was subsequently adopted in 2006. She is a member of the high-level committee that prepared the first draft of the Emigration and Overseas Employment law and is on the Board of Directors of the Migrant Welfare Bank recently established by the Government of Bangladesh to provide migration loans. In the past she has served as Chair of the Asia Pacific Migration Research Network and the South Asia Migration Resource Network.

CAMBODIA

Nyda Chhinh is a PhD candidate at the School of Environment, Flinders University. His research interests are climate change adaptation and rural development and vulnerability to climate variability and change in rural Cambodia. He is a Lecturer in Environmental Economics and the Principles of Economics at the Royal University of Phnom Penh (RUPP). His current research projects include 'Building Capacity to Adapt to Climate Change in Southeast Asia' and a feasibility study, 'Payment of Forest Environmental Services in Cambodia', in collaboration with Hue University, Vietnam. Previous research projects include 'Costs and Benefits Analysis of Small Scale Jatropha Curcas Plantation in Cambodia' and 'Impacts of ENSO on Rice Production in Cambodia'.

CHINA

Joanna I. Lewis is Assistant Professor of Science, Technology and International Affairs at Georgetown University's Edmund A. Walsh School of Foreign Service. Her research focuses on energy and environmental issues in China, including renewable energy industry development and climate change policy. She has published many journal articles, book chapters and reports and her book, *Green Innovation in China: China's Wind Power Industry and the Global Transition to a Low-carbon Economy* was published in 2013. Lewis serves as international Advisor to the Energy Foundation China Sustainable Energy Program in Beijing. She has worked for several governmental, non-governmental and international organizations and is Lead Author of the Intergovernmental Panel on Climate Change's Fifth Assessment Report. She holds a PhD in Energy and Resources from the University of California, Berkeley and a BA in Environmental Science and Policy from Duke University.

HONG KONG

Jolene Lin is Associate Professor of Law at the University of Hong Kong. Her research interests are climate change law and policy, environmental law and transnational legal theory. Her recent work focuses on biofuels governance, see, for example, 'Governing biofuels: a principal-agent analysis of the European Union biofuels certification regime and the Clean Development Mechanism', *Journal of Environmental Law*, 2012. She is an associate member of the Asia-Pacific Centre for Environmental Law and an editor of *Transnational Environmental Law*.

INDIA

Jawed Usmani is an Officer of the Indian Administrative Service, currently the Chief Secretary of Uttar Pradesh with responsibility for the coordination, monitoring and control of the functioning of all departments and agencies of the state government. He has more than three decades of experience in Indian administration at various levels, in the districts, the state government and at the centre. He has served as District Magistrate in two districts, Registrar of a central university, Managing Director of a state public sector undertaking, Minister of Economic Cooperation at the Indian Mission in Kathmandu, Secretary to the Chief Minister of Uttar Pradesh and Director and Joint Secretary in the Prime Minister's Office. Usmani has worked at the World Bank and spent a sabbatical year with Boston University and the Center for the Advanced Study of India, University of Pennsylvania, where he was a Visiting Scholar conducting research on climate change and its implications for India. Usmani has an MBA from the Indian Institute of Management, Ahmedabad and an MSc in Social Policy and Planning from the London School of Economics.

Sushil Vachani is Professor of Strategy and Innovation at Boston University. He has served as Special Assistant (for the India Initiative) to the University's President, Faculty Director of the Doctoral Program at the School of Management, Chair of the Strategy and Policy Department and Faculty Director of the International Management Program, Japan. He previously worked with the Boston Consulting Group, where he designed business strategies for American, Japanese and European multinationals. Vachani has also worked with Philips India, the Tata Administrative Service and Tata Motors in India. He is on the Board of Trustees of the Deshpande Foundation. His research interests include climate change, strategy and innovation at the bottom of the pyramid, multinational–government relations, the impact of non-governmental organizations on international business and management of diversified multinationals. His research has been published in the *Journal of International Business Studies*, *International Business Review*, *Harvard Business Review*, *California Management Review* and other publications. He is editor of *Transformations in Global Governance: Implications for Multinationals and Other Stakeholders* (2006), co-editor of *The Role of MNCs in Global Poverty Reduction* (2006) and author of *Multinationals in India: Strategic Product Choices* (1991). Vachani received his doctorate in International Business from Harvard Business School, a Postgraduate Diploma in Management from the Indian Institute of Management, Ahmedabad and a Bachelor of Technology degree from the Indian Institute of Technology, Kanpur.

NEPAL

Bhaskar Singh Karky is a Resource Economist at the International Centre for Integrated Mountain Development (ICIMOD). He was Coordinator for the Reducing Emissions from Deforestation and Forest Degradation (REDD) project for India and Nepal. He has participated in United Nations Framework Convention on Climate Change negotiations and follows climate policy developments. Karky has worked with the National Trust for Nature Conservation, the Centre for Micro-Finance and Danida in Nepal. He has published several papers on climate policy, REDD+, payment for environmental services, micro-finance and renewable energy. He received his PhD in the Economics of Climate Change Policy from the University of Twente, an MSc in Agricultural Development Economics from Reading University and a BSc in Agriculture from the University of Western Sydney.

SINGAPORE

Sofiah Jamil is a PhD candidate at the Australian National University and Adjunct Research Associate at the Centre for Non-Traditional Security (NTS) Studies at the S. Rajaratnam School of International Studies (RSIS), Nanyang Technological University (NTU). She was previously Associate

Research Fellow at the Centre for NTS Studies, where she co-led two programmes, 'Climate Change, Environmental Security and Natural Disasters' and 'Energy Security'. Jamil regularly contributes to the Centre for NTS Studies' publications, such as the *NTS Insight* and *NTS Bulletin*. Other publications include book chapters such as 'Beyond food for fuel: the little red dot in GCC-ASEAN relations' in *Asia-Gulf Economic Relations in the 21st Century: The Local to Global Transformation* (2013); 'Energy and non-traditional security in East Asia' and 'China's energy efficiency policies' in *Energy and Non-traditional Security in Asia* (2012). Jamil serves on the Board of Management of the Young Association of Muslim Professionals in Singapore. Her research interests are contemporary Muslim politics, human security and environmental issues. She has a BA Honours in Political Science and International Relations from the University of Western Australia and an MSc in International Relations from RSIS at NTU.

SOUTH KOREA

Hye-Jin Jung is Research Professor at the Asian Institute for Energy, Environment and Sustainability. He has published articles and participates in research projects on sustainable urban planning and design in response to climate change, greenhouse gas (GHG) emission reduction and mitigation policy and the building of a GHG emission inventory. Jung works to improve comprehensive GHG management systems and GHG reduction from university campuses in Korea and has been responsible for the establishment and operation of the Seoul National University Greenhouse Gas and Energy Management Center. He has started to focus on developing teenage education programmes on climate change to provide both knowledge and practice for future leaders. Jung received his PhD in Urban Design from the Graduate School of Environmental Studies, Seoul National University.

Chankook Kim is Assistant Professor in the Department of Environmental Education at the Korea National University of Education. He is also a Fellow of the Asian Institute for Energy, Environment and Sustainability in Seoul. His research areas lie at the intersection of environmental studies and education, including climate change education, education for sustainability and environmental communication on climate change. In the area of climate change education, Chankook Kim's research interests are how citizens understand scientific information on climate change and how educators deal with the issue of climate change. He received his PhD in Human Dimensions of Environment and Natural Resources from Ohio State University.

Ki-Ho Kim is presently the founding Director of the Research Institute for Climate Change Response (RICCR). He is the former Director of the Asian Institute for Energy, Environment & Sustainability (AIEES). He is Professor Emeritus of Architecture and Urban Design at the Graduate School of Environmental Studies at Seoul National University. Professor Kim has taught at Seoul National University for the past 30 years. Previously, he held various administrative positions including Dean of the Graduate School of Environmental Studies at Seoul National University, Director of the CEO Environmental Management Forum, and Director of the Environmental Planning Institute. Professor Kim earned his Bachelor of Engineering in Architecture from Seoul National University and his Master of Architecture from the University of Minnesota.

Foreword

Scientific projections are necessarily subject to scientific questioning and climate change is no exception. There are sceptics who question whether available data, including the fact that 11 of the 12 years from 1995 to 2006 rank among the 12 warmest years since 1850, really establish a long-term trend. That said, the ranks of sceptics are thinning and the dominant opinion is moving inexorably towards an acceptance of, and consequent concern with, climate change as a very real phenomenon.

The leading international body for the assessment of climate change, established by the United Nations Environment Programme and the World Meteorological Organization, is the Intergovernmental Panel on Climate Change (IPCC). The IPCC projects that the Earth's surface temperature could rise by an additional 1.8–4°C by the end of this century and that the sea level could rise by 18–59 cm. These changes will lead to a dramatic increase in the scale and frequency of major climate events such as floods, droughts, hurricanes and higher storm surges, all of which will have devastating consequences for the people impacted. Recent examples of destruction caused by such extreme events are the flash floods in Uttarakhand, India in 2013 and Hurricane Sandy in the USA in 2012. They may or may not be evidence that climate change is underway but certainly demonstrate the enormous loss that can be inflicted on human life and economic value.

These considerations suggest that it is necessary to take bold and comprehensive steps to deal with this challenge. The world must react to the challenge in two ways: it must take steps to mitigate carbon emissions and thus reduce the extent of climate change and it must take steps to adapt to such climate change as is unavoidable. The pursuit of mitigation strategies runs into the problem of externality. The benefits of mitigation action are not restricted to the country that takes mitigation steps. They extend to the whole world, and in the face of this externality, it is well established that individual countries will do much less than they should. A successful mitigation strategy therefore requires international cooperation in which all countries join collectively. For this to happen, the burden of mitigation has to be equitably distributed across all countries. The ongoing United Nations Framework Convention on Climate Change (UNFCCC) negotiations were expected to come up with an international agreement on a fair distribution of the burden of the costs of mitigation. It is unfortunate that very little progress has been made thus far. Developing countries argue that since it is the industrialized countries that are responsible for the historical accumulation of greenhouse gases, and also since they are economically stronger, they must bear the principal burden. The world has yet to agree on what is a fair outcome and pending that there is no agreement on mitigation.

Unlike mitigation, which involves externality, adaptation does not. The benefits of adaptation actions accrue to the country taking them and it is therefore important to define adaptation strategies that should be followed, given that some climate change seems unavoidable. This book provides a timely and thoughtful discussion of strategies for adaptation to climate change, which can complement mitigation strategies being developed by other experts throughout the world, to reduce the risk of disaster.

The book focuses on eight countries in Asia, which together are home to 2.8 billion people, about 40 per cent of the world's population. They are among the areas that will experience the most severe impact of climate change and many are extremely vulnerable in terms of capacity to cope with the impact. These territories have very diverse geographics (extensive coastlines, deserts, mountains and megacities), demographics (population ranging from less than ten million to over a billion) and economics (variation in size of the economy as well as per capita income).

This diversity provides the authors with an opportunity to discuss adaptation strategies for many of the urgent and critical threats expected to be faced by sectors that are most impacted by climate change. Their conclusions are relevant beyond the individual countries analysed. Specifically, they discuss the challenges for large cities due to water scarcity, flooding, sea level rise and heat island effect; the dangers to biodiversity due to degradation of fragile environments; the devastating implications for human health due to rise in diseases such as malaria, dengue, cholera and diarrhoea; the potential large-scale food shortages due to rural water scarcity, declining agricultural yields and disruption of food imports; and the migration of large numbers due to disruption of livelihoods in rural areas.

The authors discuss adaptation strategies for each of these significant threats. They highlight plans that have been successful and explore what else can be done to address the gaps. Their observations and recommendations should prove very useful to policy makers in other countries facing similar threats who are focusing on building their adaptive capacity to cope with climate change. An important point made in this volume is that though effective policies and systems at a national level are critical to meet the challenges faced by a country, these must be complemented at the local level by the private sector, local government and civil society. The success, or otherwise, of the strategies highlighted in this book have lessons for key constituencies at all levels and for how to achieve coordination among them.

The authors recognize the trade-offs inherent in the desire for economic growth now versus the need to invest in building adaptive capacity for the future, and urge the governments to actively engage in careful evaluation of those trade-offs. Their research highlights where such trade-offs have been most successful and offer lessons for national and local decision makers. They also stress the need for action now, for leveraging technology and for continuing an iterative process of research, action and monitoring to refine and develop enhanced strategies. These recommendations can inform the agendas for research institutes and government agencies that plan and implement adaptation strategies.

The authors in this book have been deeply engaged in the area of the environment, energy and climate change, as academics at leading universities and as members of research institutes or government bodies, and they draw on their enormous knowledge and experience to present valuable discussion of adaptation to climate change in Asia.

An aspect that is not specifically quantified is the cost of adaptation. Many of the measures proposed will involve significant costs. As long as the costs are less than the cost of not adapting, there is an economic case for undertaking adaptation but this does involve a draft on available resources. To the extent to which this burden is being imposed on developing countries because of action predominantly taken by industrialized countries over a long period, there is a moral case for the industrialized countries sharing this burden, especially from low incomes countries. As in the case of mitigation, this is one of the items on the agenda of the UNFCCC negotiations, and is effectively mired with all the others. Nonetheless, as the authors emphasize, this is the time for action on many fronts and I urge the policy makers to give serious attention to the strategies presented in this book.

Montek Singh Ahluwalia
Deputy Chairman, Planning Commission
Government of India

1. Adaptation to climate change in Asia

Sushil Vachani and Jawed Usmani

The frequency and scale of damage inflicted by climate-related disasters, such as floods, drought, heat waves and hurricanes, has been increasing at an alarming rate. As Margareta Wahlström, chief of the United Nations International Strategy for Disaster Reduction, notes, 'Climate change is accelerating the pace and intensity of extreme weather events'[1] (Wahlström 2012, p. 1), causing significant loss of life and economic damage (CRED 2013b). Asia bears a disproportionately large share of these losses and the brunt of the suffering is borne by the poorest people. Though earthquakes caused over half of the losses from natural disasters between 2002 and 2011, a significant share (31 percent of deaths and 42 percent of economic damage) resulted from climate-related disasters, creating a sense of urgency for adaptation to climate change (CRED 2013a).

Fortunately, the world is finally recognizing the critical need for adaptation to climate change, which presents immense challenges for human health and livelihood, food and water security, biodiversity and preservation of cities. It is imperative that countries succeed in designing and implementing effective adaptation strategies because failure to do so will leave them at risk of great human suffering and enormous economic loss.

Previously, supporters of mitigation feared that stressing adaptation would empower opponents of mitigation who argued that human ingenuity at adaptation would neutralize the threat of global warming. In addition, pressure to comply with agreements to reduce greenhouse gas emissions encouraged preferential allocation of resources for mitigation rather than adaptation (Wreford et al. 2010). Over time, however, there has been greater recognition that adaptation and mitigation are both critically important to contain the detrimental effects of climate change (Klein et al. 2007). Strategies to address climate change must carefully integrate policies for mitigation and adaptation rather than follow either one in isolation, which makes it important to clearly understand the adaptation needs

1

and strategies being adopted in different regions (Ingham et al. 2013; Kane and Yohe 2000).

This book provides readers with a broad overview of challenges presented by climate change in Asia and the adaptation strategies being implemented. The choice of territories allows us to cover different areas in which action is urgently needed. Some of the territories are among those most threatened by climate change.[2] Four of them appear in Wheeler's (2011) ranking of the top 20 territories facing risk from extreme weather in 2015: China (ranked 1), India (3), Bangladesh (8) and Hong Kong (13).[3] Together, the territories featured in this book give us the opportunity to reflect on the gravest threats presented by climate change in Asia, and learn from their strategies to ensure food and water supply, address urban problems, support migrant workers, protect coastal cities and preserve biodiversity. The examination of their adaptation strategies also highlights the immensity of the implementation task, which requires widespread outreach to build resilient communities with the assistance of civil society. Other countries can learn important lessons from the experts writing about the challenges faced and strategies being adopted.

1.1   EFFECT OF CLIMATE CHANGE

The authoritative source on climate change is the Intergovernmental Panel on Climate Change (IPCC). In its most recent appraisal of climate change, completed as part of the preparation for the Fifth Assessment Report, due to be released later in 2014, the IPCC indicated that the Earth's surface temperature rose 0.85°C between 1880 and 2012 (Stocker et al. 2013). Projections based on varying scenarios for rise in carbon dioxide level indicate that towards the end of this century (2081–2100) global surface temperature will be between 0.3 to 4.8°C higher than it was between 1986–2005. This projected warming will lead to more extreme climate events, such as violent storms or drought, which will cause greater damage to homes and livelihoods, compromise food production and reduce availability of fresh water. Rising temperature will also present many challenges to human health and biodiversity. Even an increase of 1.5 to 2.5°C presents a higher risk of extinction for 20 to 30 percent of the Earth's animal and plant species.[4]

Over the last century the average sea level rose 0.19 meters (Stocker et al. 2013). Projections indicate that sea level is likely to be between 0.26 and 0.82 meters higher in 2081–2100 than it was in 1986–2005. Higher sea level, together with more violent storms, presents a grave threat of damage to coastal cities, such as Hurricane Sandy inflicted on New York City and coastal New Jersey in 2012.

1.1.1 Effect on Asia

In its 2007 assessment, the IPCC identified the following areas in which Asia would be affected by climate change.[5]

Fresh water scarcity
It is estimated that by the middle of the twenty-first century around a billion people, spread across several regions, will face declining fresh water availability. As glaciers melt, they will initially supply more water to rivers, increasing floods. Over time, however, these glaciers will shrink to the extent that the much-needed water they supply to rivers during the dry summer months will be inadequate to support the large population downstream.

Food shortage
Some Asian countries will face great challenges in feeding their people as agricultural yields decline significantly in certain regions, while rising elsewhere. The adverse impact, driven primarily by higher temperature and water scarcity, will vary considerably across provinces and districts within countries. Action is needed not only to attenuate the damage in adversely affected regions but also secure benefits where climate change could potentially enhance output.

Harm to human health
More extreme climate events, such as heat waves, flood and drought, will exact a toll on human health as more people succumb to heat and dehydration, and diseases such as malaria, dengue, cholera and diarrhea. The effects will be devastating for the large Asian population that lives in extreme poverty and has little access to health care.

Threat to coastal populations
Major coastal cities will face greater threat from the combination of more violent storms and higher sea level. Farming communities will lose agricultural output as a result of salt intrusion in the soil and groundwater. Most coastal cities are poorly equipped to deal with this onslaught of nature, though some, such as Shanghai, have begun to bolster defenses.

Economic loss

The effects mentioned above will translate into significant economic hardship in rural and urban areas alike. For business centers like Singapore, Hong Kong, Seoul, Shanghai and Mumbai, the disruption of infrastructure and communications caused by violent storms will result in serious economic loss that ripples through the economy. In rural areas, climate change will wreak havoc on the large share of the population dependent on agriculture. As climate change damages the fragile mountain environments of countries like Nepal it will reduce economic value derived from tourism and the generation of hydro-electric power.

1.1.2   Anthropogenic Climate Change

This book does not address the question of whether climate change is caused by humans. It relies on assessments made by credible sources, such as the IPCC and governments of major countries, that climate change is indeed occurring, and restricts its focus to examining projected impact and adaptation strategies. However, given that there is some debate on whether climate change is indeed caused by human activity, we refer to a recent study that addresses the question. Cook et al. (2013) presented a comprehensive review of research papers that focus on climate change and concluded 'the number of papers rejecting the consensus on AGW [anthropogenic global warming] is a vanishingly small proportion of the published research' (Cook et al. 2013, p. 1). They found that among researchers who have written about climate change and express an opinion on whether its causes are anthropogenic, over 97 percent are convinced that it is caused by human actions.

1.2   ADAPTATION

Adaptation to climate change refers to adjustments made to limit damage or enhance benefit from climate change and its effects, as indicated in the definition provided by the IPCC (Box 1.1). Adaptation might involve changes in a variety of systems as suggested by the United Nations Framework Convention on Climate Change (UNFCCC). These changes could involve social development (for example, creation of social networks to build resilience), technical practices (for example, agricultural techniques), institutions (for example, support systems for migrants), infrastructure (for example, physical barriers to contain flood) or market mechanisms (for example, water pricing).

BOX 1.1   DEFINITIONS OF ADAPTATION

From the Intergovernmental Panel on Climate Change (IPCC)

> Adaptation: In human systems, the process of adjustment to actual or expected climate and its effects, in order to moderate harm or exploit beneficial opportunities. In natural systems, the process of adjustment to actual climate and its effects; human intervention may facilitate adjustment to expected climate. (Field et al. 2012, p.5)

From the United Nations Framework Convention on Climate Change (UNFCCC)

> Adaptation refers to adjustments in ecological, social, or economic systems in response to actual or expected climatic stimuli and their effects or impacts. It refers to changes in processes, practices, and structures to moderate potential damages or to benefit from opportunities associated with climate change.

Source: http://unfccc.int/focus/adaptation/items/6999.php (accessed 27 April 2013).

1.2.1   Geographic Coverage

This book focuses on adaptation to climate change in eight Asian territories, seven countries – Bangladesh, Cambodia, China, India, Nepal, Singapore and South Korea – and Hong Kong, which is a special administrative region within China. The territories vary in their physical, demographic and economic characteristics, which affect the nature of the challenges they face from climate change as well as their ability to address them.

The territories have diverse physical characteristics. Some have great exposure to the sea, which makes them vulnerable to destruction from the combined effect of violent storms and higher sea level. Singapore and Hong Kong consist of several islands and China, India, Bangladesh and South Korea also have extensive coastlines. Inadequate infrastructure creates daunting logistical challenges in many of these countries (Bangladesh, Cambodia, China, India and Nepal), making it difficult to access rural areas devastated by extreme climate events. About two-thirds of Nepal consists of mountains, where the environment is fragile and especially susceptible to damage by climate change.

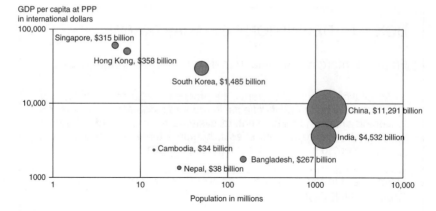

Note: The size of each circle corresponds to the GDP in 2011 at purchasing price parity (PPP) in current international dollars, which is also noted next to the territory's name.

Source: Based on data obtained from the World Development Indicators, World Bank Database, available at: http://databank.worldbank.org/data/views/variableSelection/ selectvariables.aspx?source=world-development-indicators# (accessed April 28, 2013).

Figure 1.1   Economic and demographic profiles, 2011

Together, these territories are home to 2.8 billion people, about 40 percent of the world's population and nearly 70 percent of Asia's. Their population ranges from less than ten million for Singapore and Hong Kong, to over a billion for China and India. Some are highly urbanized (the share of the population living in cities is 90 percent for South Korea and 100 percent for Hong Kong and Singapore), while others are fairly rural – less than a third of the population of India, Bangladesh, Cambodia and Nepal (and about half of China's population) lives in cities.[6] Urban areas are prone to a unique set of problems arising from climate change.

China and India are among the five largest economies in the world.[7] However, they are not the richest economies. Singapore, Hong Kong and South Korea have the highest per capita income and Nepal, Bangladesh and Cambodia the lowest (Figure 1.1). Poorer nations will find it difficult to muster adequate resources to adapt to climate change, and the poorest segments of their population stand to suffer the most.

The economies vary considerably in structure (Figure 1.2). Cambodia and Nepal are agrarian societies with agriculture accounting for about 36 percent of GDP for each. While services have grown significantly in Bangladesh and India (contributing 53 to 54 percent of GDP), agriculture remains important as it contributes around 18 percent, and also because a large proportion of the population is rural and subsists on very low

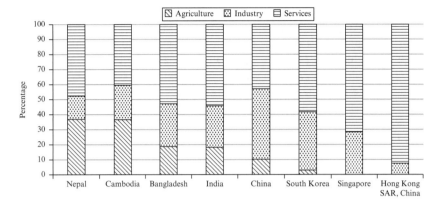

Source: Based on data obtained from the World Development Indicators, World Bank
Database, available at http://databank.worldbank.org/data/views/variableSelection/
selectvariables.aspx?source=world-development-indicators# (accessed 28 April 2013).

Figure 1.2   *Share of agriculture, industry and services in GDP of the*
               country or region, 2010

income, derived from agriculture. Heavy reliance on agriculture makes
these economies especially vulnerable to the vagaries of climate change.
Even in China, where agriculture accounts for just 10 percent of GDP,
since half the population is rural, adverse impact on agriculture causes
great hardship.

Industry contributes far more to the economies of China (47 percent of
GDP) and South Korea (39 percent) than the others. With the development of global value chains, industries are highly susceptible to disruption from climate-related disasters. For example, in 2011, severe floods in
Thailand halted manufacturing of key components, such as disk drives
and automobile parts, disrupting global supply chains across continents
(Ralph 2012; Soble 2011).

Hong Kong and Singapore are advanced service-based economies with
93 percent and 72 percent of GDP contributed by services, respectively.
South Korea (59 percent of GDP) and China (43 percent) also rely a good
deal on services. As economies shift from agriculture to manufacturing
and services, the rural population drifts into cities, straining facilities,
causing overcrowding, creating social tension and increasing vulnerability
to extreme climate events. This challenge will only worsen as Asia develops
and urbanization soars.[8]

1.3   THEMES

This book includes a chapter on each of the eight chosen territories. The authors highlight areas of adaptation they have identified as important for the country or region covered. Despite the unique focus of each of the chapters, there are themes that cut across many of them in varying degrees. These are briefly outlined here.

1.3.1   Urban Challenges

Since 2010, more than half the world's population is living in cities. Already straining to provide essential services to residents, cities are exposed to a range of threats from climate change. These include excessive heat, water scarcity, floods and higher sea level, as discussed in the chapters on Singapore, Hong Kong and South Korea.

Urban heat island effect
Urban populations in many Asian cities are reeling under the effect of high summer temperatures. The damage caused by rising temperature is exacerbated by the urban heat island effect, which refers to higher temperature in built-up areas compared to those with vegetation. As Sofiah Jamil writes in her chapter on Singapore, the adverse effects of climate change, such as higher temperature and rainfall, are 'amplified by the process of urbanization.' Temperature in 'highly urbanized areas' is 4°C higher than in 'well-planted' areas. The burden placed by heat waves is disproportionately high for the poor and infirm at the bottom of the pyramid, who lack the means to protect themselves.

Governments have implemented significant measures to tackle the urban heat island effect. Jamil explains how Singapore has provided incentives for construction companies to develop energy-efficient buildings. She notes that instead of requiring greenery on the ground the government has sought to create 'green spaces . . . above ground level, such as green roofs and vertical planting to create green walls.' Jolene Lin discusses how Hong Kong has addressed its urban heat challenges by enacting laws to enforce adaptation guidelines, for example, design guidelines for provision of open spaces, landscaping, insulation and air circulation.

Ki-Ho Kim, Hye-Jin Jung and Chankook Kim observe that South Korea's urban heat problem has worsened as the share of its population living in cities has risen significantly in the past 50 years. Between 1974 and 1997, average temperature rose 1.5°C in cities, compared with 0.58°C in the countryside and coast. One of the major causes for this difference in temperature rise in cities and other areas is the urban heat island effect, which has accounted for 30 to 40 percent of the rise in temperature in South Korean cities.

In response to the heat challenge, Seoul's city government and civil society have collaborated to create more green spaces, for example, by building greenways, city forests and the World Cup Park on sites that were previously built-up areas. Regulations require developers to set aside green space in new developments. The authors summarize the ten principles for building climate-proof cities developed by Ki-Ho Kim (the chapter's lead author) in previous work with Kook-Hyun Moon. They identify the importance of building design to ensure accessibility of green public spaces for pedestrians, and laying out buildings of different height within a development so as to optimize availability of natural light across dwellings and reduce energy consumption. South Korea's achievements in implementing urban planning to create greener urban environments are impressive and could serve as examples for other countries.

Urban water scarcity
The water supply in many Asian cities is already stretched so far that residents have to survive on meager allocations, which will shrink further with climate change. Jolene Lin observes that rainfall variability will worsen water shortage during lean periods in Hong Kong. The city's water storage capacity is inadequate and it will be forced to increase reliance on import from mainland China. Singapore probably provides the best example of planning for water shortages over the next 50 years as it reduces dependence on water imported from Malaysia. It plans to increase the contribution of water obtained by recycling and desalination from 40 percent to 80 percent, while retaining more rainfall in expanded catchment areas.

Urban floods
One of the big problems facing cities is a greater risk of floods, which threaten homes, commercial property, business activity and livelihoods. Cities face greater risk of floods because higher rainfall that arrives in intense bursts can quickly overwhelm drainage systems. Kim, Jung and Kim note that Seoul has become more prone to floods over the last century, as there are four times as many days with heavy rainfall now. Precipitation has intensified as well. In Seoul, in 2011, it rose as high as 588 millimeters in a three-day period. Over time, the area covered by buildings has grown rapidly, shrinking the surface through which rainwater can percolate underground.

Sofiah Jamil commends the achievements of Singapore's government in developing the economy and raising its people's living standards. She also acknowledges the significant strides made in flood protection; for example, by developing the drainage system, areas susceptible to flooding have been dramatically reduced even though the country's land mass has grown by well over a third owing to reclamation. However, she is concerned that rapid urbanization has left the country inadequately prepared to address challenges posed by climate change.

Jolene Lin notes that Hong Kong received its highest recorded rainfall (1.34 meters in a month) in June 2008, with 30 centimeters falling in a single day and inflicting damage worth $75 million. She describes how Hong Kong improved flood protection by consolidating authority within one agency and developing a strategy that sets standards, monitors, maintains and upgrades drainage systems to meet those standards, and uses laws to protect water courses.

Hong Kong's approach can be useful for other cities; often the actions needed for effectively combating the challenge of urban flooding through proper zoning and development of drainage systems that are appropriately maintained require close coordination among different government agencies.

Sea level rise
The damage inflicted by storms on coastal cities is increasing with the rise in sea level. Jolene Lin points out that even if the frequency and intensity of cyclones that bring heavy rain to Hong Kong remain unchanged, their damage to the city will increase as higher sea levels result in bigger storm surges.

Joanna Lewis discusses the imminent danger the rising sea poses for Chinese coastal cities, including Shanghai, where vulnerability to flood is exacerbated by the fact that the city is sinking as a result of excess groundwater withdrawal and the weight of skyscrapers that is causing the soil to subside. Lewis feels that Shanghai's four-tier defense system to protect it from the sea and the Huangpu River, including walls, levees, water gates and drainage and discharge systems, could serve as a model for other coastal cities. Given the large number of coastal cities at risk from climate-induced sea level rise, there is a lot that can be learned from Shanghai's strategy.

1.3.2   Biodiversity

Climate change presents significant risk to biodiversity. Bhaskar Karky cautions that Nepal's exceptionally rich flora and fauna, already endangered by human encroachment, are under stress from climate change, which also threatens extinction of endangered species, such as the snow leopard.

Climate change has increased the risk to ecosystems in both Hong Kong and Singapore, where biodiversity has already been hurt by urbanization. Jamil observes that since Singapore's independence about 50 years ago, it has suffered significant loss of biodiversity as land was cleared or reclaimed. It has lost over 95 percent of its original mangroves, '67% of its birds, 40% of its animals and 5% of its amphibians and reptiles, and . . . 39% of its native coastal plants' (Biodiversity Portal of Singapore 2010).

Lewis notes that China, with its 'long history of destructive deforestation practices,' is legislating protection of forests and developing action plans to preserve them in the face of threats from climate change. China is focusing attention on measures to restrict logging, prevent fires and shield forests from pests and disease. It has also instituted programs to preserve wetlands and marine ecosystems.

Many countries have experienced environmental damage and loss of biodiversity as a result of urbanization and industrialization, exacerbated by climate change. This creates opportunities to learn from the mistakes and successes of China, Nepal, Hong Kong and Singapore as discussed in the following chapters.

1.3.3 Health

Climate change will increase risk to human health in a number of ways. In addition to threatening lives with heat waves, higher temperatures will aid transmission of deadly disease in previously unaffected areas. Floods will also worsen the spread of disease. In India, the government expects that rise in temperature will create conditions for malaria to be transmitted to northern regions where it was previously absent, while increasing the period during which it appears in southern parts of the country. However, at some locations where it has occurred year round, higher temperatures will reduce the transmission period because the temperature will be too high for the disease-carrying mosquito to survive. The incidence of several other diseases will also rise, making it important for India to develop stronger public health systems and effective means for early detection and containment of epidemics.

Lin observes that Hong Kong learned from the unfortunate experience of the 2003 severe acute respiratory syndrome (SARS) epidemic, which spread rapidly and caused panic because the government was unprepared. The experience led to the institution of a comprehensive plan to monitor and contain the spread of disease, backed by laws to enable enforcement. Jamil points to studies that show that diseases like malaria and dengue have spread as a result of a rise in temperature. Over the years, Singapore has made remarkable progress in providing a sanitary environment by pre-empting the creation of urban slums, which are all too common as cities grow. It has achieved this by passing and implementing laws ensuring public hygiene.

The experiences of Hong Kong and Singapore can be helpful for other countries in designing and implementing strategies to control the spread of deadly diseases, which is inevitable with climate change.

1.3.4   Food

Climate change presents a significant threat to countries' ability to feed their people by compromising agricultural output and disrupting imports from adversely affected regions.

Agriculture

For many developing countries the biggest threat from climate change is that of disruption of agriculture, which supports a significant segment of their population. Cambodia is affected by floods and drought practically every year and suffers extensive damage to its paddy cultivation, as discussed by Nyda Chhinh. The flood in 2000, the worst in 70 years, killed 100 people and inflicted $170 million worth of damage on agriculture. In 2011, crops were lost on more than 10 percent of Cambodia's land devoted to paddy (rice). These problems will become worse as the country faces an increase in the intensity and frequency of natural hazards.

Chhinh fears climate change will worsen flood and drought, magnifying the hardship for farmers, most of whom are poor and cultivate paddy on small farms with traditional practices, paying steep rent to landlords. They cannot afford to invest in irrigation or purchase adequate fertilizer, with the result that their productivity is only two-thirds of farmers in neighboring Vietnam and Thailand. They are forced to borrow from moneylenders and sell their produce below market price. Their marginal existence leads Chhinh to surmise, 'It is evident that the effects of climate change are tangible and beyond the coping range of farmers. Policy development that will enhance the adaptive capacity of farmers to natural hazards, especially flood and drought that are worsening due to the changing climate, is urgently required.'

The Cambodian government's policy response to protect farmers from flood and drought relies on agricultural extension services to enhance their capacity to address challenges, supplemented with easier credit access to enable them to increase investment in improving productivity. Chhinh discusses Cambodia's institutional arrangements for coordinating these responses, which include collaboration across government agencies. An example of the output of inter-ministerial collaboration is the 'Strategy for

Agriculture and Water, 2006–2010,' which proposes tackling drought and flood by 'improving water resources, irrigation and land management, and improving agricultural and water research, education and extension.' The government also relies on mechanisms to foster community involvement, for example, working with Farmer Water User Committees.

Sushil Vachani and Jawed Usmani describe changes in the rainfall pattern as a major threat to Indian agriculture. Though overall rainfall will increase, it will vary significantly across seasons and regions. Greater rainfall intensity will increase floods and damage crops. Decrease in rainfall between December and February will hurt winter crops. About 16 percent of India's agricultural land is vulnerable to drought, which causes significant damage to agricultural output, especially for certain crops, like potato.

An important part of the Indian government's adaptation strategy is to promote robust crops that can withstand flood and drought, and encourage practices to protect them; for example, planting tomatoes and onions on raised beds to save them from floods. The government plans to promote conservation agriculture, which avoids mechanical disturbance of soil and provides it with organic cover derived from previous harvests.

India's yield of wheat and rice is estimated to be just over a third of potential output, and varies significantly across regions. A large share of the population in states with low yield lives in poverty. The adverse impact of climate change on productivity will be felt more heavily by these poor farmers, most of whom cultivate crops on rain-fed land.

The productivity challenges faced by Indian agriculture are common in many agrarian societies where lack of knowledge about best practices, and inability to invest in productivity owing to poverty and inadequate institutional support at the bottom of the pyramid, compromise adaptability (Vachani and Smith 2008). This creates a need for massive outreach at the local level to strengthen communities, which is discussed later in this chapter.

Rural water scarcity
Agricultural output is closely linked to availability of water. Many Asian countries depend on water from rivers that originate in the Himalayas. In his chapter on Nepal, Karky reminds us that as Himalayan glaciers melt, the water they provide Asian countries during the dry summer months preceding the monsoon will diminish and hurt agriculture. As in many developing countries, Nepalese farmers are poor and subsist on small plots of rain-fed land, which makes them especially vulnerable to extreme climate events and water scarcity. The country's ability to store water is inadequate to provide sufficient buffer for the anticipated water shortage.

Both China and India have to make do with much less water per person than the world's average. Lewis observes that in some parts of China where water is already scarce, its availability is projected to decrease dramatically. In the basin of the Huang, Huai and Hai rivers (referred to as the 3H basin), water flow has fallen 15 to 41 percent over the last 20 years. The government is safeguarding agriculture in the region by increasing land under irrigation, creating awareness of threats and building local capacity to respond to scarcity. Lewis notes that China's attempts to boost agricultural resilience through water conservation, promotion of climate-resistant seeds and crop diversification have raised grain production in drought-prone areas and increased water productivity by 10 to 30 percent. There is great interest in developing special varieties of rice that can tolerate drought, given its prominence in the Chinese diet, and the fact that its cultivation consumes 50 to 70 percent of the country's water.

One of the factors hurting water conservation is the popular perception of water as 'inexhaustible,' which is difficult to change without pricing that discourages usage. Lewis notes that the Chinese government is trying to ensure efficient water consumption with measures such as permits that help control usage. The Chinese government is also trying to deal with water pollution. It has introduced regulations to monitor and ensure quality of drinking water in each province. Lewis wonders, however, if the government will succeed with implementation and voices concern about vulnerability of some of the infrastructure projects.

Per capita water availability in India is less than a quarter of the world's average. Since a large proportion of water is consumed in agriculture, there is little room to divert it from other uses, making India's agricultural output extremely dependent on overall water supply. As climate change reduces water availability, Vachani and Usmani express concern that the effect on agriculture could be devastating. The Indian government expects 'crisis-like' conditions with regard to water availability in most regions. Agriculture will also be hurt by the rise in sea level as it contaminates coastal soil and groundwater with salt.

About 40 percent of India's potentially usable water supply comes from groundwater. It is imperative for the country to clearly understand the pattern of consumption and recharge of groundwater, and discourage its overuse. The Central Groundwater Authority regulates use of groundwater in areas where it is overexploited. There is, however, a need for more stringent and comprehensive measures to discourage water consumption, for example, with appropriate pricing, which is difficult to impose politically, a challenge faced by many countries.

Vulnerability of food imports

Regions that meet their food needs with imports are exposed to the disruptive effects of climate change on agricultural production at their sources. Jamil describes Singapore's high dependence on imported food as a matter of great concern for the country. In 2008, 90 percent of Singapore's food was sourced from 31 countries. Climate-related disruption of food production in those countries could affect availability. In 2011, Singapore's imports of major commodities from Australia were significantly disrupted by floods in Queensland, demonstrating the vulnerability of its supply chain. Other countries can learn much from Singapore's strategies for securing food supply outlined by Jamil; for example, setting up farms abroad, encouraging local production, research investments and managing demand through measures such as waste reduction.

1.3.5 Migration

Extreme weather events cause severe hardship for people living in rural areas, disrupting livelihoods and forcing many to migrate. In their chapter on Bangladesh, Tasneem Siddiqui and Motasim Billah develop a case for changing the negative perception of migration as a 'consequence of failure to adapt,' which leads to strategies for decreasing its scope. Bangladesh's policy response to climate change has either ignored migration or, in some instances, such as the National Adaptation Program of Action, characterized it as 'an undesirable outcome of climate change and associated it with increased incidence of crime.' Siddiqui and Billah see migration differently – as building the resilience of affected families and communities. They sense the government has gradually begun to acknowledge the magnitude of the challenge – in 2009, the Bangladesh Climate Change Strategy and Action Plan forecast that as many as 20 million Bangladeshis could be displaced, calling for more vigorous human resource development so that migrants could obtain employment in foreign markets. Still, some of the key institutions that oversee programs for migrants, for example, the ministries for labor and expatriate welfare, are not directly involved with climate change adaptation programs. Civil society, which provides significant support for affected people, also seems to undervalue migration as an adaptation strategy.

Siddiqui and Billah report on their study of migration from three districts of Bangladesh that are periodically affected by severe climate events such as cyclones, storm surges, riverbank erosion and drought. They observe that environmental changes have forced communities to change professions, and migration has helped diversify opportunities. Migration has been facilitated by the development of communications, transport infrastructure and networking with labor and construction contractors in urban destinations. Over time, migrants benefit from a 'process of step migration,' where initial exposure and experience leads to greater domestic opportunities and, eventually, international migration. Siddiqui and Billah feel that if migration is perceived as an adaptive strategy, there might be greater focus on training migrants for higher-paying work in the formal sector, which would be a step up from low-paying informal sector jobs. It would help to enhance engagement of government ministries that deal with training, as well as institutions that can help migrants fund relocation.

As the *Financial Times* columnist, Martin Wolf, puts it, given that adaptation to climate change 'may well include large-scale movements of people' (Wolf 2013, p. 9), it is time governments acknowledged the inevitable and began systematically planning for it. Countries could learn from the experience of Bangladesh, where policy orientation has evolved substantially to be more supportive of migrants. They must be proactive in addressing possible ethnic tensions resulting from migration across provinces, which Lewis refers to in the case of China, where climate-induced water scarcity forces people to move.

Migration is a matter of concern not just for home countries, from where migrants pull up roots, but also host countries, where they settle. Host countries could learn from Singapore, which is attempting to ensure the welfare of immigrants as some of them fall upon hard times, as described by Jamil.

1.3.6 Implementation

The analyses of different territories presented by the book's contributors indicate that success of adaptation strategies critically depends on effective implementation, which requires clarity and political commitment at senior levels of government, aggressive outreach at local levels to strengthen communities by engaging civil society and leveraging technology to magnify the impact of implementation strategies.

Clarity and political commitment

Effectiveness of adaptation depends on how the country's government and its people perceive the threat from climate change, and how determined they are to implement change. Lack of conviction about the threat at senior levels of government, ambivalence toward adaptation and inability to signal its importance can compromise adaptation. Lin views the Hong Kong administration as lacking 'political will to do more than the bare minimum in a reactive fashion,' with the result that its efforts fall short of countries like Singapore and South Korea. She notes that the government's 'passive approach' stems from an attitude 'prioritizing economic goals over environmental concerns' that is rooted in its 'historical and political context,' which allowed it to use its special status as a region within China to remain free of the more stringent demands placed on developed countries to reduce greenhouse gas emissions. This is exacerbated by 'institutional inertia' resulting from poor relations between the government's Chief Executive and the bureaucracy. Unfortunately, this prevents Hong Kong from fully leveraging its 'wealth and human talent' to address climate change challenges.

Lewis indicates that 'real questions remain among [Chinese] government leaders about the true extent of vulnerability' to climate change and that 'better characterization of the risks they face would allow policy makers to better understand where climate adaptation strategies should rank in terms of other policy priorities.' However, she reports that the leadership's receptivity to scientific assessments appears to be increasing as evident from the inclusion of adaptation measures in recent policies.

In his assessment of Cambodia's preparedness for climate change, Chhinh detects 'overall adaptive weakness' that 'requires improvement in government intervention, which is still at an early stage in terms of geographic coverage and practical options.'

Countries can experience a dramatic change in their perception of the need to address climate change and implement adaptation measures. Kim, Jung and Kim outline the circumstances that led to the impressive turnaround in South Korea's attitude toward preservation of the environment. When it was ranked close to the bottom of 142 countries on the Environmental Sustainability Index in 2001, the nation was shocked, and this led to significant measures to address the challenges posed by global warming. By 2012, South Korea was a 'strong performer' on the Environmental Performance Index (EPI) (which had replaced the Environmental Sustainability Index by then).[9]

Aggressive outreach to strengthen communities by engaging civil society

Several authors stress the importance of dedicated outreach to create awareness, strengthen communities and stimulate action at the local level, especially through collaboration with civil society.

Success depends on effective implementation at the local level. As Kim, Jung and Kim assert, 'in the war against climate change, victory depends on the policies of the local or city government,' and that cross-sectoral partnerships are instrumental in the effort: '[d]iverse societal sectors, with civil society at the center, at both the national and municipal levels have collaborated to improve the environmental situation in South Korea.'

Siddiqui and Billah describe the high degree of involvement of

Bangladeshi non-governmental organizations (NGOs) in implementing community-based programs to evaluate vulnerability and share knowledge about climate-resilient agriculture and risk-reduction strategies, and building adaptive capacity. Chhinh stresses the need to provide 'training on how to build institutional, technical and capital resilience for climate-proof irrigation' in Cambodia. As part of its water mission, India has set a target for improving efficiency of water use by 20 percent. Vachani and Usmani describe the requirement of a careful assessment of the impact of climate change together with strategies to conserve and augment water resources implemented through government programs and involvement of civil society.

Jamil feels that civil society could help improve Singapore's preparedness for addressing the adverse effects of climate change. She acknowledges that Singapore's 'meticulous urban planning' has given it a strong base of resources and capabilities to deal with the adverse impact of climate change, but that its efforts need to be supported with community engagement. She suggests that in order to develop a community that is resilient enough to address climate change Singapore needs to reinforce 'adaptive capacities relating to social capital and community competence' in addition to the strong adaptive capacity 'related to economic development, and information and communication.'

Jamil discusses how social resilience is difficult to achieve in cities owing to a sense of independence and a lower commitment to community and attachment to the environment. Many Singaporeans are unaware of the challenges presented by climate change, such as the rise in sea level, and so are unprepared for these events. They appear to suffer from a 'sense of infallibility,' which is combined with the sense that it is the government's responsibility, rather than their own, to prepare for disasters. Jamil sees the need to foster a 'culture of interdependence' in Singapore by building 'multi-sectoral collaborations' that focus on 'the needs and capacities of people, rather than the needs and capacities of infrastructure.' She feels that as part of this effort it is important to map collaborative efforts and take measures to ensure sustainability of ad hoc initiatives.

Capacity building An important theme cutting across chapters is the need to bolster adaptive capacity. Siddiqui and Billah point to the need for strengthening the ability of Bangladeshi government agencies to assist people who choose to migrate by providing them with training and other assistance. Karky concludes that Nepal needs to strengthen its human resources for development planning and recommends blending climate change adaptation strategies into mainstream development planning.

In some countries there is a high level of expertise at senior levels of government, but inadequate field capability. Jamil's assessment is that Singapore has developed 'advanced top-down technical and infrastructural capabilities for climate change adaptation (CCA) measures,' but needs to strengthen 'bottom-up adaptive capacities and consciousness on climate change needed to complement government efforts.' She discusses many ways to build capacity, including formal course work. She suggests one way to prepare Singaporeans for disasters is to engage them in relief work in neighboring countries, such as the Philippines, Thailand and Indonesia, when disasters strike the region. This would, of course, require careful preparation and supervision.

India is implementing its adaptation programs through the National Action Plan on Climate Change, which comprises eight missions focusing on different areas. Vachani and Usmani note that the success of its adaptation strategy for agriculture hinges on the effectiveness of programs to reach out to millions of farmers to sensitize them to upcoming challenges, and build their capacity to deploy effective agricultural techniques. This enormous implementation task can be made more manageable through collaboration with India's large and vibrant civil society.

Coordination among key constituencies   Effective implementation, which succeeds in reaching deep into the bottom of the pyramid to provide knowledge and disseminate best practices widely, requires close coordination among diverse sets of players. In Bangladesh, for example, Siddiqui and Billah recommend that key ministries that provide training should expand their efforts in areas from where villagers migrate. Government agencies that focus on activities affected by climate change, such as the Ministries for Labor and for Expatriates' Welfare and Overseas Employment, need to engage with bodies overseeing the implementation of adaptation programs. Karky identifies the need to give adaptation greater emphasis by bringing it into important mainstream programs through 'dovetail[ing the] development process with adaptation activity at the local level' in Nepal.

Leveraging technology to magnify impact

Technology can be a powerful enabling factor in implementing adaptation strategies, as demonstrated by South Korea. Kim, Jung and Kim discuss how South Korea uses modern technology to provide effective fine-grained, real-time warnings about hazards such as fires. Taking advantage of the country's high connectivity (80 percent of households), the government is using sophisticated technology-based services to provide the public with real-time information relating to disaster relief, health, agriculture and forestry. In the event of a forest fire, for example, the system can promptly deliver information about the extent of the threat, its forecast for expansion and suggested countermeasures to the public as well as first responders so that they can ensure the public's safety.

The South Korean experience with advanced systems and technologies can be useful in other countries; for example, in addressing the need expressed by Chhinh for 'mapping flood and drought-prone localities' and implementing systems to provide hazard warnings to farmers in Cambodia. As penetration of smart phones and Internet connectivity rise in China and India, these technologies could be deployed to provide high-quality information and early warnings to large segments of their populations. As Lewis notes, China is devoting resources to develop and improve systems to provide early warning of extreme climate events and infectious diseases. It intends to integrate systems across provinces into a national network to facilitate monitoring and reporting.

Countries can also gain from the experience of Hong Kong, which underscores the importance of effective organization. Lin observes that 'Hong Kong has an excellent warning system for bad weather conditions' and a system for tackling emergencies that clearly defines tasks and responsibilities for different organizational units, and delegates authority to field units. Hong Kong uses the media to provide alerts about health hazards to the public as well as agencies that provide community support.

1.4   CONCLUSIONS

There is clearly an enormous amount of planning and implementation needed at all levels of society to adapt to climate change. Several authors surmise that the adaptation efforts taken by governments are at early stages. In order to ensure effectiveness, adaptation strategies must be informed by careful research so that resources are deployed where they deliver maximum benefit. However, given the seriousness of the threats presented by climate change, it is important that implementation of adaptation strategies be infused with a sense of urgency. Given resource scarcity, trade-offs are inevitable when setting priorities. In the past, adaptation to climate change was often placed behind other objectives, especially economic development. It is now time to give it greater importance not just at national levels, but globally. Effective adaptation requires recognition of global interdependence – effects of climate change easily cut across from one part of the world to another and solutions require coordination across countries.

1.4.1   Research

The enormous adaptation task needs to be informed by careful research in a number of areas. Based on their experience in Bangladesh, Siddiqui and Billah point to the importance of studying how people use migration to adapt to changes arising from climate change as well as a range of other factors. Karky recommends that the impact of climate change on Nepalese agriculture be studied more closely using fine-grained local data. Chhinh considers it important for Cambodia to test agricultural practices that would improve yield during drought and flood, and develop varieties of seeds that can withstand climate stress.

Lewis describes the Chinese government's effort to promote the development of robust cultivars that can withstand climate stress and threat from pests and disease. The government claims that grain production has risen 40 percent with the use of such varieties on 95 percent of agricultural land. Lewis observes that Chinese institutions have improved their ability to monitor climate change, but it is important for the country to further enhance its scientific capability so that it can remain at the forefront for predicting the impact of climate change.

Vachani and Usmani describe the Indian government's support for a network of public research institutions that focus on developing varieties of crops that can thrive as conditions evolve with climate change. The Indian government stresses the importance of sustaining research as it may take years to obtain results. Given the serious impact that water shortage will have on Indian agriculture, it is very important for the government to assess the impact of climate change on water resources across regions and identify specific areas for action. As climate change reduces water flow into rivers from the Himalayas, downstream countries will need to evaluate variation in micro-environments across regions to determine the specific effects on different rivers.

Lin points out that studies are needed to assess how different sectors of the Hong Kong economy will be affected so that adaptation can be tailored accordingly. She also recommends examination of the specific effects on the health of different vulnerable segments such as the poor, older and sick residents. These would supplement measures that Hong Kong has already put in place to study infectious diseases after the major scare from the SARS epidemic in 2003.

1.4.2   Trade-offs

An important consideration in formulating and implementing adaptation strategies is creating awareness of trade-offs inherent in them. Often, other priorities supersede adaptation. In the case of South Korea, Kim, Jung and Kim explain that the reason why Seoul has only about half the Food and Agriculture Organization-recommended 9 square meters of residential parks per person is that during the country's period of rapid development, the government placed a higher priority on meeting housing demand than providing open spaces. While commending China's adaptation programs, Lewis remarks that 'the task is vast, and rapid economic development predominantly trumps thoughtful, climate-resilient development, particularly when resources are scarce.' Jamil notes that in Singapore, preservation of biodiversity has tended to get lower priority than measures to secure water resources and protect against floods.

1.4.3 Global Interdependence

Several authors discuss the risk of disruption of global value chains as a result of extreme climate events. Given the interdependencies created by globalization, the adverse effects of climate change are likely to ripple across countries, with implications for the flow of food, water, industrial components, information and people. Hong Kong imports significant quantities of food from Thailand and China, Nepal depends on food from India and Singapore procures food from several countries. When supplies are disrupted, prices can rise sharply, causing severe hardship for the poorer segments even in rich societies like Singapore.

Both Hong Kong and Singapore import water, and need to worry about ensuring future supplies. Hong Kong relies on nuclear power generated in mainland China, which runs the risk of disruption. Climate change can also provide opportunities for some sectors of the economy; for example, Hong Kong-based insurance companies could avail of new business opportunities as countries attempt to manage risk arising from climate change.

Cross-border information sharing and coordination of emergency measures are critical to prevent the spread of disease. Lin notes that following the SARS epidemic, Hong Kong, Macau and China have ratcheted up cooperation to prevent the spread of epidemics. As cross-border migration increases in response to adverse effects of climate change, programs are needed to create hospitable conditions for migrants and assist them in finding livelihoods in their new homes.

The following chapters, written by experts who have specialized knowledge about adaptation to climate change in the territories covered, provide more detailed discussion of the themes summarized here.

NOTES

1. Margareta Wahlström's official title is United Nations Special Representative of the Secretary-General for Disaster Risk Reduction. Her summary assessment is supported by historical data in the International Disaster Database maintained by the Centre for Research on the Epidemiology of Disasters (CRED) at the Université Catholique de Louvain (UCL) in Brussels. See http://www.emdat.be/disaster-trends (accessed 28 August 2013).
2. We use the term territory rather than country when referring to the set of areas covered since they are not all countries. Hong Kong is a special administrative region within China, while the other seven are countries – Bangladesh, Cambodia, China, India, Nepal, Singapore and South Korea.
3. Wheeler ranks countries on climate vulnerability based on a measure of the probability of an individual residing in the country being affected by a climate-related disaster in a given year, using an econometric model built with data drawn from the EM-DAT database maintained by CRED. The model regresses vulnerability on level of atmospheric carbon as well as income, urbanization and governance factors pertaining to the country.
4. This information is drawn from the UNFCCC available at http://unfccc.int/essential_background/the_science/items/6064.php (accessed 18 June 2013). It is based on the UNFCCC's Fourth Assessment Report. The Fifth Assessment Report is expected to be released after this book is submitted for publication.
5. Based on information provided by the UNFCCC, available at http://unfccc.int/essential_background/the_science/climate_change_impacts_map/items/6225.php (accessed 18 June 2013).
6. The percentage of the population living in urban areas in 2012 for the eight territories was 17.3 percent for Nepal, 20.2 percent for Cambodia, 28.9 percent for Bangladesh, 31.7 percent for India, 51.8 percent for China, 83.5 percent for South Korea and 100 percent each for Hong Kong and Singapore. Data obtained from World Development Indicators, World Bank Database, available at http://databank.worldbank.org/data/views/variable selection/selectvariables.aspx#c_b (accessed 30 August 2013). We have presented the higher percentage of 90 percent for South Korea based on data published in the *Yearly Statistics of Land, Transport and Maritime Affairs* (2012), Korean Ministry of Land, Transport and Maritime Affairs, p. 157.
7. This is based on gross domestic product (GDP) at purchasing price parity, which adjusts nominal GDP to remove price differences across countries. They are among the top ten if one does not adjust for prices.
8. The countries with significant rural populations are experiencing rapid urbanization. The average annual rate of change of the urban population between 2010 and 2015 is projected to be 3.62 percent for Nepal, 2.96 percent for Bangladesh, 2.85 percent for China, 2.47 percent for India and 2.13 percent for Cambodia. Data obtained from CIA World Factbook website, available at https://www.cia.gov/library/publications/the-world-factbook/fields/2212.html (accessed 1 August 2013).
9. South Korea was ranked 43rd among 132 countries, as indicated in the EPI rankings available at http://epi.yale.edu/epi2012/rankings (accessed 3 June 2013).

REFERENCES

Biodiversity Portal of Singapore (2010), 'Singapore, an interesting case study', *The Diversity of Life on Earth: From Heritage to Extinction*, available at http://www.biodiversity.sg/assets/Uploads/TheDiversityOfLifeOnEarth-Chapter8.pdf (accessed 16 June 2013).

Cook, J., D. Nuccitelli, S.A. Green et al. (2013), 'Quantifying the consensus on anthropogenic global warming in the scientific literature', *Environmental Research Letters*, 1–7, available at http://iopscience.iop.org/1748–9326/8/2/024024/pdf/1748–9326_8_2_024024.pdf (accessed 22 May 2013).

CRED (Centre for Research on the Epidemiology of Disasters) (2013a), 'Disaster data: a balanced perspective', No. 30, January.

CRED (Centre for Research on the Epidemiology of Disasters) (2013b), 'Disaster data: a balanced perspective', No. 31, March.

Field, C.B., V. Barros, T.F. Stocker et al. (eds) (2012), *Managing the Risks of Extreme Events and Disasters to Advance Climate Change Adaptation*, Special Report of Working Groups I and II of the Intergovernmental Panel on Climate Change (IPCC), Cambridge and New York: Cambridge University Press.

Ingham, A., J. Ma and A.M. Ulph (2013), 'Can adaptation and mitigation be complements?', *Climatic Change*, **120**, 39–53.

Kane, S. and G. Yohe (2000), 'Societal adaptation to climate variability and change: an introduction', *Climatic Change*, **45**, 1–4.

Klein, R.J.T., S. Huq, F. Denton et al. (2007), 'Inter-relationships between adaptation and mitigation', in M.L. Parry, O.F. Canziani, J.P. Palutikof, P.J. van der Linden and C.E. Hanson (eds), *Climate Change 2007: Impacts, Adaptation and Vulnerability. Contribution of Working Group II to the Fourth Assessment Report of the Intergovernmental Panel on Climate Change*, Cambridge: Cambridge University Press, pp. 745–77.

Ralph, A. (2012), 'Pace picks up again after a year of living dangerously', *The Times*, 15 November, p. 61.

Soble, J. (2011), 'Japan carmakers hit by Thai floods', *Financial Times*, 13 October, Asia edn, p. 19.

Stocker, T.F., D. Qin and G.-K. Plattner et al. (2013), *Climate Change 2013, The Physical Science Basis: Summary for Policy Makers. Working Group I Contribution to the Fifth Assessment Report of the Intergovernmental Panel on Climate Change*, Switzerland: Intergovernmental Panel on Climate Change.

Vachani, S. and N.C. Smith (2008), 'Socially responsible distribution: strategies for reaching the bottom of the pyramid', *California Management Review*, **50** (2), 52–84.

Wahlström, M. (2012), 'Foreword', in United Nations International Strategy for Disaster Reduction (UNISDR) (ed.), *Annual report 2012*, Geneva: UNISDR, p. 1.

Wheeler, D. (2011), 'Quantifying vulnerability to climate change: implications for adaptation assistance', CGD Working Paper 240, Center for Global Development, Washington, DC, available at http://www.cgdev.org/content/publications/detail/1424759 (accessed 29 April 2013).

Wolf, M. (2013), 'Climate sceptics have already won', *Financial Times*, 22 May, p. 9.

Wreford, A., D. Moran and N. Adger (2010), *Climate Change and Agriculture: Impacts, Adaptation and Mitigation*, Paris: OECD Publishing.

2. Vulnerable and lagging behind: the case of Hong Kong

Jolene Lin

2.1 INTRODUCTION

Hong Kong is situated at the mouth of the Pearl River Delta (PRD), formed by the Xijiang (West), Beijiang (North), Dongjiang (East) and Zhujiang (Pearl) Rivers as they enter the South China Sea.[1] The territory of Hong Kong consists of four major islands and about 260 outlying islands. Hong Kong's total land area is 1104 square kilometres with a total coastline of approximately 730 kilometres. About 25 per cent of Hong Kong's land area is developed; country parks and nature reserves account for 40 per cent of Hong Kong's land use. Victoria Harbour, one of the world's busiest deep water harbours, lies between Hong Kong Island and the Kowloon Peninsula (HKSAR Government 2012a). Large urban centres are located in low-lying areas, which renders the city vulnerable to sea level rise and floods. Hong Kong's population stood at approximately seven million in 2011. It is one of the most densely populated places in the world, with a population density of 6540 people per square kilometre (HKSAR Government 2012a). Hong Kong is a wealthy city – the gross national income in 2011 was approximately US$34,000 per capita – but it also has the greatest degree of income inequality in all of Asia (Vidal 2010).[2]

Like all cities, Hong Kong is vulnerable to climate change because of the concentration of people, built infrastructure and assets in a small area. In terms of its geography, 'the physical location of Hong Kong, surrounded by water, on traditional typhoon tracks and with a dense urban setting, makes it particularly vulnerable to climate change' (Welford 2009, p.3). The city's dependence on imported food, water, energy and other products to meet its population's basic needs adds to its vulnerability to the impacts of climate change.

Unlike governance in the West, bureaucracies still assume a central governing role in Asian countries regardless of whether they are democratically elected (Cheung and Scott 2003). A case study on Hong Kong's approach towards adaptation will therefore necessarily focus on governmental policies. This chapter reviews the Hong Kong government's current approach towards climate change adaptation and concludes that most of the policies in place are 'business as usual' such as flood management and public health surveillance. The policies may be tweaked by technocrats/civil servants to address climate change risks, giving rise to a sectoral, bureaucratic and piecemeal approach towards adaptation. This chapter argues that the government's failure to provide strategy and leadership in tackling climate change is due to a lack of political will and institutional inertia.

Section 2.2 provides a broad overview of the likely impact of climate change on Hong Kong and the city's adaptation needs. Section 2.3 reviews the policies and strategies that play a key role in Hong Kong's adaptation to climate change even though these measures were not designed with climate change in mind. Section 2.4 argues that many of these policy measures are 'business as usual' and there have been no real efforts to engage civil society, business and various sectors of society in what must necessarily be a collective effort to respond to the risks posed by climate change. It then elaborates on the reasons behind the political reluctance and institutional inertia within government that severely limits Hong Kong's ability to adapt to climate change. Section 2.5 concludes.

2.2   THE MAJOR CLIMATE CHANGE RISKS THAT HONG KONG FACES

This section provides an overview of the predicted changes in temperature, rainfall, storms and sea level in Hong Kong and the associated adaptation challenges that the city faces.

2.2.1   Temperature

Data collected by the National Climate Centre of the China Meteorological Administration from 28 monitoring stations in southern China for the period 1951–2000 shows that average temperatures in the PRD region have increased (Tracy et al. 2006). In Hong Kong, the annual mean temperature has risen by approximately 0.12°C per decade since data collection began in 1884 (Leung et al. 2006). Between 1980 and 2009, rural areas became warmer at a rate of around 0.2°C per decade. The corresponding rise in urban areas was much higher at around 0.6°C per decade and this temperature difference can be attributed to the 'urban heat island effect' (whereby urban temperatures are higher because of heat emitted from buildings and the characteristics of airflow within dense urban spaces) (Leung et al. 2004).

The Hong Kong Observatory predicts that annual mean temperatures will have risen by 3.5°C by the end of the twenty-first century. The annual number of very hot days (maximum temperature of 33°C or above) in summer will rise from 11 days to 24 days, while hot nights (minimum temperature of 28°C or above) will rise to 30 per year – four times the current normal level. At the same time, winters will become warmer too. It is predicted that the number of cold days (minimum temperature of 12°C or below) will decrease from 21 days to less than 1 day per year (Leung et al. 2006, p. 3). Temperature rises will increase the city's energy use and will also increase the probability of extreme weather events such as heavy rainstorms.

2.2.2   Rainfall

The Hong Kong Observatory predicts that average annual rainfall will increase by about 1 per cent per decade in this century and that the year-to-year variability in rainfall will increase (Leung et al. 2006, p. 16). In the years of low rainfall, there will be increased pressure on water resources that are already under significant pressure in the PRD region due to large-scale manufacturing, increased population and urban development. Hong Kong has had a long-standing problem of insufficient drinking water. Natural storage capacity in the form of reservoirs and water tanks is limited, forcing Hong Kong to import about 75 per cent of its drinking water supply from Dong Jiang in Guangdong Province (HKSAR Government 2008, p. 16). Decreased rainfall will place additional stress on limited water resources, rendering Hong Kong vulnerable to the impacts of decreased rainfall in the PRD region.

On the opposite end of the spectrum are the problems caused by heavy and prolonged rainfall. Long periods of heavy rainfall can lead to high levels of rainwater that neither evaporates nor penetrates the ground surface to become groundwater. This is known as excess run-off. Excess run-off increases the risks of flooding and landslide damage. In non-tropical cities, extreme precipitation events rarely cause casualties but frequently cause damage to property and infrastructure. Tropical cities like Hong Kong, however, have experienced situations whereby extreme rainfall has caused flash flooding and mud flows, leading to casualties and flood damage to parts of the city.[3] In June 2008, Hong Kong experienced the heaviest rainfall in recorded history with a monthly total of 1346 millimetres. The damage caused by the heavy rainfall on 7 June 2008, when 301 millimetres fell on that single day, was estimated at US$75 million.[4]

Tropical cyclones are another major cause of heavy rainfall in Hong Kong. There is much uncertainty concerning the impact of global warming on the frequency and intensity of tropical cyclone activity, and therefore no definitive answer to whether tropical cyclone activity has increased or will increase due to climate change. In November 2006, participants of the World Meteorological Organization (WMO) International Workshop on Tropical Cyclones issued a 'Statement on Tropical Cyclones and Climate Change' as requested by the WMO and many heads of national meteorological and hydrological services so that they can better advise governments on how to respond to climate change effects (WMO 2006, para. 1). Despite significant uncertainties and lack of consensus in the scientific community, there is sufficient consistency in the results of high-resolution global models and regional hurricane models to conclude that 'it is likely that some increase in tropical cyclone intensity will occur if the climate continues to warm. A robust result in model simulations of tropical cyclones in a warmer climate is that there will be an increase in precipitation associated with these systems' (WMO 2006, paras 18–19). In addition, scientists at the Hong Kong University of Science and Technology believe that typhoons will track further east as the Western Pacific warm pool becomes warmer. Consequently, typhoons will be more likely to miss Hong Kong and make landfall in Japan, which will result in more rainfall over Hong Kong and increase the risk of flooding (Leung et al. 2006, p. 18).

Even if the current level of tropical cyclone activity remains unchanged, storms and tropical cyclones can be expected to cause greater destruction because of higher storm surges associated with higher sea levels (WMO 2006, para. 25). Higher storm surges can overcome existing coastal defences and cause great economic damage and loss of life in coastal population centres like Hong Kong.

2.2.3   Sea Level Rise

Along the coast of Guangdong province, where Hong Kong is situated in the middle of that coastline, the sea level is rising at a rate of 1 centimetre each year and could rise 30 centimetres by 2030 (Leung et al. 2006, p. 20; Pauw and Francesch-Huidobro 2010, p. 107). Urbanization, development and population growth render population centres along the coastline and coastal ecosystems more vulnerable to the impacts of sea level rise. Development, for example, usually worsens soil erosion, damages wetlands and changes the amount of sediments that are delivered and deposited in the coastal areas (US Environmental Protection Agency 2013). In the United States, an example of such impacts of development can be found in coastal Louisiana. In the past few decades, as a result of human activity that has altered the sediment system of the Mississippi River, as well as oil and water extraction that have made land subside, as much as 1900 square miles of wetlands have been destroyed. The remaining wetlands are also unable to serve as natural buffers to flooding because the sediment they receive is inadequate to keep pace with the rising sea (US Climate Change Science Program 2008). A rise in the sea level will also lead to higher wave and storm surges along the coast, which can pose significant risks to property and lives. An impact that is a concern for policy-makers in many jurisdictions is the loss of fresh water as the sea level rise pushes sea water further into underground water tables.[5] In Hong Kong, saline water intrusion is not a significant concern as the city imports most of its water and there are no significant agricultural and industrial sectors.

2.2.4 Vulnerability Assessment

In December 2010, the Hong Kong government published 'A study of climate change in Hong Kong – feasibility study', a consultation paper intended to inform government policies on climate change and that included the city's first climate change vulnerability assessment (HKSAR Government Environmental Protection Department 2010). Based on a future climate scenario (in the year 2100), the assessment identified eight key sectors as highly vulnerable to climate change impacts. The criteria used to identify key vulnerabilities include the magnitude of impacts, timing of impacts, importance of the sector under study, the distributional aspects of impacts and vulnerabilities and potential for adaptation.[6] The eight key sectors are: biodiversity and nature conservation; the built environment and infrastructure; business and industry; energy supply; financial services; food resources; human health; and water resources.[7]

Biodiversity and nature conservation[8]
While there is inadequate knowledge of the full extent of Hong Kong's ecosystems and biodiversity and there is incomplete baseline data, it is well established that Hong Kong has a number of endemic species and is disproportionally rich in biodiversity. Much of this local biodiversity is already under substantial pressures from human activities such as urbanization, hill fires, pollution and the introduction of foreign species. Ecosystems (for example, coral reefs) and biodiversity are highly sensitive to climate change and, given the rate of change, autonomous adaptation is unlikely to take place. Further, despite some conservation efforts that aim to limit non-climate stress on species and habitats, there are limited realistic options for preserving many endemic species in areas that become climatically unsuitable. The process of natural selection may result in these

species being replaced by better suited or invasive species. The study concluded that increasing loss of biodiversity and colonization of invasive species are expected, and these could occur on an annual basis.

Built environment and infrastructure[9]
The physical infrastructure of a city, including its buildings and transportation networks, has low potential for adaptation. For example, there is little room for adapting existing development to the risk of loss of land in low-lying or reclaimed areas as sea levels rise. The substantial uncertainties surrounding the magnitude and rate of change in future sea levels make it difficult to formulate appropriate adaptation measures even if it is possible to use modelling and imaging technologies to identify the infrastructure that is most at risk. Some groups such as the less wealthy and small enterprises will be more susceptible to the impacts of climate change on the built environment as they cannot afford to relocate or have less capacity to recover from major events. As rainfall patterns change, prolonged periods of heavy rainfall will place additional stress on Hong Kong's drainage and sewerage infrastructure. More frequent and severe rainstorms will significantly increase the number and scale of rain-induced landslides on both natural hillsides and man-made slopes. This will pose greater danger to public safety. Compared to buildings and transportation infrastructure, communication infrastructure has greater potential for adaptation as it is relatively less extensive and the costs of adaptation are likely to be lower than the costs of adapting buildings and highways.

Business and industry[10]
It is not possible to generalize the impacts of climate change on the business and industry sector as it covers a wide range of activities from financial services to shipping. Each sector will be exposed to different climate change risks and opportunities, rendering it necessary to carry out more specialized investigation into each sub-sector. This sector is sensitive to the impacts of climate change on other places because of globalization. For example, a retailer in Hong Kong will be affected by impacts along the supply trade, on trade partners and financial markets, and disruption in transport or communication infrastructure. Over the past decades, manufacturing activities have moved across the border from Hong Kong to the mainland, particularly the PRD economic zone.

> Hong Kong has been the source of approximately two-thirds of the cumulative foreign direct investment in the [PRD] since 1979. In addition to the tens of thousands of small and medium-sized Hong Kong firms active in the Pearl River Delta region are several large players, such as Hutchison in port services,

VTech in electronics, Hopewell in highways, Jardines in retailing, HSBC and Bank of East Asia in banking, China Light & Power in power generation, and several Hong Kong developers in property and hotels. (Enright et al., Scott & Associates 2005, p. 8)

The impacts of climate change on this area will have a significant impact on Hong Kong's business sector, and it should be noted that resource and technical constraints will make it more difficult for the manufacturing sector in the PRD to cope with climate-related disturbances or extreme events.

Energy supply[11]
Climate change is expected to change Hong Kong's energy demand patterns. For example, more electricity will be needed to provide air-conditioning for people to cope with hotter days in summer. This would challenge the power supply infrastructure that may be affected by extreme weather events. At the same time, Hong Kong is committed to transiting from its heavy reliance on fossil fuels towards low-carbon energy options. Electricity generation is responsible for 67 per cent of greenhouse gas emissions in Hong Kong (HKSAR Government Environment Bureau 2010, p. 34). In 2009, about half of the city's electricity was generated from coal.[12] In the government's climate change strategy consultation document, it was proposed that by 2020, the fuel mix for power generation in Hong Kong would be as follows: no more than 10 per cent from coal, approximately 40 per cent from natural gas, approximately 3–4 per cent from renewable energy such as wind and solar power, and approximately 50 per cent from nuclear power.[13] From an adaptation perspective, more knowledge about how climate change could impact the supply chain of primary fuel imports and supply is needed to guide policy-makers. For example, given that Hong Kong's nuclear power is imported from the Daya Bay power plant in Shenzhen, any climate change-related factors that could affect the safety and operation of the plant will be relevant to Hong Kong. Close cooperation between the relevant bureaus in Hong Kong and Shenzhen will be necessary to address these risks and safety concerns.

Financial services[14]
The financial services sector plays a very important role in Hong Kong's economy. It employs 5.3 per cent of the city's workforce and contributes 16 per cent of the city's gross domestic product (GDP).[15] As a highly globalized industry, this sector (particularly the insurance and banking businesses) will suffer from sudden and unexpected climate-related events overseas. However, quite apart from climate risks, there are opportunities

that climate change presents to this sector. For example, the insurance industry is expected to play a significant role in climate risk management and many insurers and re-insurers are already exploring potential business opportunities. Larger corporations, particularly the multinational ones, are more aware of the need to 'climate proof' their business operations and assets because of environmental reporting requirements imposed by their shareholders, securities law or undertaken voluntarily.[16] Other corporations have limited preparedness, which could have severe repercussions on Hong Kong's economy.

Food resources[17]

Hong Kong is not self-sufficient in its food supply and relies heavily on imports from a few trading partners, notably Thailand (for rice) and mainland China. It is therefore vulnerable to supply disruptions caused by climate change impacts on food-supplying countries. Food security, especially for the poor and elderly, will be threatened when food prices soar because crops are damaged by extreme weather events. In August 2012, the worst drought in more than 50 years in the United States caused the corn harvest to drop 13 per cent from the year before (Abbott 2012; Meyer et al. 2012). Global corn prices surged more than 60 per cent while global soy prices also soared because supply was affected by drought in South America. Such events will affect places like Hong Kong as corn and soy are used in a range of food products as well as feedstock for animal farming. Food producers will pass on the costs of higher corn prices to their customers.

Human health[18]

The principal concerns about the human health effects of climate change include 'injuries and fatalities related to severe weather events and heat waves; infectious diseases related to changes in vector biology, water, and food contamination' and 'respiratory and cardiovascular disease related to worsening air pollution', which is already a significant problem in Hong Kong (Frumkin et al. 2008, p. 435). The government's climate change study suggests that the impacts of communicable diseases (such as vector-borne diseases) are likely to be reduced due to increased preventative, monitoring, surveillance and reactive measures.[19] Vulnerable groups such as the elderly, sick and the poor are likely to be disproportionally affected but more investigation is required to understand how these groups will be impacted.

Water resources[20]

Although the projections show increased annual rainfall in the future, it is difficult to project how the annual rainfall distribution might change.

The change in distribution could have significant repercussions for water security in Hong Kong. As mentioned earlier, Hong Kong depends heavily on imported water from Dongjiang to meet its needs. This dependence renders it vulnerable to fluctuations in rainfall supply and distribution patterns in the greater PRD region. Current policy in Hong Kong stresses demand management to ensure that demand is kept within supply capacity.

2.3   CURRENT POLICIES AND STRATEGIES

Unlike other major cities like London, Hong Kong does not have an adaptation strategy that has been developed specifically to address climate change.[21] Instead, Hong Kong's approach is to rely on existing policy and technical measures, enhanced in certain cases to take climate change into account, to respond to its adaptation needs and challenges. For example, as flooding has historically been a challenge for the city, Hong Kong has a well-developed flood anticipation and prevention system. Policy-makers believe that this system, with some tweaking, will be able to adequately enhance Hong Kong's adaptive capacity. While this is the case for flood risks, it is not necessarily the case for other vulnerable sectors. The stock-taking exercise that culminated in 'A study of climate change in Hong Kong – feasibility study' revealed that few sectors believe that their policies build adaptive capacity for climate change as a by-product of other strategy aims. Further, apart from the public health, built environment and water resources sectors, none of the other sectors have actively considered the issue of climate change adaptation.[22] It can be argued that these attitudes reflect a combination of complacency and overconfidence in the relatively high level of adaptive capacity that a developed society like Hong Kong possesses.[23] Complacency is, as described by O'Brien et al. (2006), a lack of awareness of potential dangers and an accompanying self-satisfaction that no action is needed to adapt to climate change. Complacency is generated when, amongst other things, the focus of the community's climate change discourse is primarily on mitigation options, with adaptation appearing as an afterthought emphasizing technological solutions. In Hong Kong, the nascent and limited climate change discourse is certainly mitigation-centric and the government, as the key provider of public goods, can do more in terms of working with stakeholders such as businesses, tertiary institutions and community networks to raise awareness of the impacts of climate change, dispel the general air of complacency about adaptation and thereby enhance the city's adaptive capacity. The rest of this section presents a review of current policies and strategies that play a key role in Hong Kong's adaptation to climate change even though these measures were not designed with climate change in mind.

2.3.1   Responding to Extreme Temperatures

It is not unusual, especially during summer, for the outdoor temperature to soar above 31°C. The high temperature is exacerbated by the 'urban heat island effect' in highly built-up areas such as Central, Causeway Bay and Mongkok (HKSAR Government 2007, p. 99). As mentioned earlier, the impacts of such high temperatures include increased energy consumption, greater air pollution (higher air temperatures promote the formation of ground-level ozone) and negative health effects. It is not possible to prevent heat waves, but it is possible to mitigate the 'urban heat island effect' through effective urban planning and design. This includes increasing tree and vegetative cover, creating 'green' roofs and using 'cool pavements'; that is, pavements built with 'materials that reflect more solar energy, enhance water evaporation or have been otherwise modified to remain cooler than conventional pavements' (US Environmental Protection Agency 2012, para. 1). The *Hong Kong Planning Standards and Guidelines* is a comprehensive set of urban design guidelines concerning a range of issues such as air ventilation, the creation of open spaces and landscaping (HKSAR Government Planning Department 2011). There are, however, no laws that require new and redevelopment projects to address the 'urban heat island effect'.

It is common to use air-conditioning to keep buildings cool in Hong Kong. However, as pointed out in the government's climate change consultation report, 'this solution is unsustainable as it is energy intensive, contributes to greenhouse gas emissions and the waste heat generated can exacerbate the urban heat island effect'.[24] Unfortunately, while there exists a Code of Practice for Overall Thermal Transfer Value in Buildings that provides technical guidance on the reduction of heat transfer through the building envelope (in particular, the external walls and roofs of a commercial building), this Code of Practice does not apply to private developments.[25] Further, buildings in Hong Kong are usually poorly insulated. There is therefore much room for improving the energy efficiency of Hong Kong's building stock.

Heat stress and cold stress may result from periods of extreme temperatures. A recent study on individual help-seeking behaviour during periods of elevated temperatures in Hong Kong indicates that the social groups most vulnerable to the adverse impacts of heat stress are those who are over 75 years of age, women, those who are not married, those living in densely populated areas such as Kowloon and the homeless (Chan et al.

2011). Presently, when unusually high or low temperatures are anticipated, the government's Department of Health will issue health warnings that are communicated to the public via the media. These health warnings also serve to activate relevant government agencies and social welfare charity groups into providing community outreach services and emergency relief such as opening temporary shelters and distributing blankets. The Department of Health also seeks to provide health guidance on coping with extreme temperatures through radio and television shows, the Internet and talks at community centres and homes for the elderly.[26]

2.3.2 Responding to Extreme Weather Events

Hong Kong has an excellent warning system for bad weather conditions. When severe weather conditions are predicted to affect Hong Kong, the Observatory issues early warnings so that the public is able to take precautionary measures.[27] In addition, the government has developed an emergency response system that clearly sets out how relevant departments and organizations will respond to any man-made or natural emergency situation that threatens life, property and public security. The system is designed to handle the three main phases of any emergency response – rescue, recovery and restoration – and is based on the policy of 'keep[ing] the emergency response as simple as possible by (a) limiting the number of involved departments and agencies; (b) limiting the levels of communication within the emergency response system; and (c) delegating necessary authority and responsibility to those at the scene of an emergency' (HKSAR Government Security Bureau 2012, chapter 3, para. 3.4). In addition to the emergency response system, there are a number of contingency plans that address specific security risks. Relevant to climate change adaptation are the contingency plan for Natural Disasters and the Daya Bay contingency plan (which provides information about the government's preparedness and response to nuclear emergencies).[28]

2.3.3 Responding to Diseases

The severe acute respiratory syndrome (SARS) epidemic in 2003 inflicted severe social, humanitarian and economic costs on Hong Kong.[29] The SARS outbreak reached epidemic proportions so quickly and explosively because of the lack of preparedness on the part of the Department of Health and the Hospital Authority (Lee 2003). Inadequate communication also caused waves of panic through the community and weakened cooperation and support from the public (Lee 2003). The SARS epidemic eventually turned out to be an important lesson. It led to the development of a clear emergency response plan in the face of public health emergencies and unprecedented cross-border cooperation with mainland China and Macau in disease control.[30] Significant resources have also been devoted to increasing Hong Kong's ability to respond effectively to infectious diseases. For example, new research centres devoted to the study of emerging infectious diseases have been established.[31]

A new legal framework has also been put in place. The Prevention and Control of Disease Ordinance (Cap. 599) and the Prevention and Control of Disease Regulation (Cap. 599A) came into operation on 14 July 2008. These two pieces of legislation replace the Quarantine and Prevention of Disease Ordinance (Cap. 141). The new laws were introduced to bring Hong Kong's obsolete disease prevention and control framework in line with international standards set out in the 2005 International Health Regulations of the World Health Organization (WHO).[32] The new regulatory regime includes disease monitoring systems that focus on detecting diseases at the border checkpoints, a clear set of operating procedures and control measures to reduce the spread of diseases across borders and the systematic collection of epidemiological data by the Department of Health to inform public health strategies.

Other preventative measures that the government takes to avoid disease outbreaks include public education, awareness raising and information sharing among health professionals and childhood vaccinations. The Department of Health also maintains close contact with the Food and Environmental Hygiene Department to obtain climate-related vector-based data and food safety monitoring information.[33]

2.3.4 Responding to Flood Risks

Prone to floods, Hong Kong saw the implementation of flood protection measures soon after she came under British rule in 1841 (Pauw and Francesch-Huidobro 2010, p. 109). The earliest drainage system was constructed mainly for sanitary reasons. Shortly after, the drainage (flood protection) systems and sewer systems were clearly separated. In 1987, the Director of Civil Engineering assumed responsibility for coordinating all aspects of flood control in Hong Kong, rectifying a situation of fragmented authority across various agencies and departments. In 1990, the first comprehensive drainage and flood control strategy was adopted and the Drainage Services Department was established.

The key features of Hong Kong's flood prevention strategy are: (1) the use of flood protection standards for new drainage works and for gradual improvement to existing drainage systems; (2) comprehensive and regular monitoring to identify new drainage works and improvement to existing drainage systems; (3) constantly bringing the capacity of new and existing drainage systems in line with the flood protection standards as far as practicable; (4) use of legal powers under the Land Drainage Ordinance (Cap. 446) to protect major water courses, especially those within privately held land; and (5) vigilant maintenance of the stormwater drainage system (HKSAR Government Drainage Services Department 2011). Flood protection standards are a crucial aspect of any flood control and prevention strategy. The Return Period of an event refers to the average time interval (in years) between which the event is expected to happen. In Hong Kong, urban drainage trunk systems are required to be able to withstand flood events of a 200-year Return Period, urban drainage branch systems to withstand flood events of a 50-year Return Period and village drainage to withstand flood events of a ten-year Return Period (HKSAR Government Drainage Services Department 2011). The Drainage Services Department also operates a 24-hour telephone hotline for reporting flood cases. Finally, since 1990, the Public Works Departments has followed guidelines that require planners of government infrastructure projects to assume that the rate of mean sea level rise stands at 10 millimetres per year in order to address the potential impacts of climate change (HKSAR Government Environmental Protection Department 2010, p. 55).

2.4 BARRIERS TO EFFECTIVE ADAPTATION

As can be seen from Section 2.3, the Hong Kong government does not lack technical knowledge or resources to take up the issue of climate change. Instead, what it lacks is political will to do more than the bare minimum in a reactive fashion. The government's thinking on climate change has been described as 'decades behind' and its efforts pale in comparison to those of regional peers including Singapore and South Korea (Chu and Schroeder 2010, p. 299). It can be said that '[c]limate change has entered the political rhetoric, but has not yet triggered any meaningful action' (Chu and Schroeder 2010, p. 291). For example, there are few if any community-based adaptation initiatives in Hong Kong. Community-based adaptation may be defined as 'a process focused on those communities that are most vulnerable to climate change, based on the premise of understanding how climate change will affect the local environment and a community's assets and capacities' (Ensor and Berger 2009, p. 231). It is premised on the belief that societies resist change that is perceived to be imposed by 'outsiders' and is therefore deeply rooted in the local context. Community-based adaptation 'requires those working with communities to engage with indigenous capacities, knowledge and practices of coping with past and present climate-related hazards' (Ensor and Berger 2009, p. 231). In Hong Kong, such engagement with local communities such as farmers is minimal.

There are two main reasons for the government's passive approach. First, the historical and political context in which Hong Kong participates in the international climate change treaty regime both reflects and reinforces the government's attitude of prioritizing economic goals over environmental concerns. The lack of political will to mitigate greenhouse gas (GHG) emissions has crippled the development of a broader policy discourse that will include adaptation. Second, in addition to a lack of political will, there is a high degree of institutional inertia within government due to the poor working relationship between the Chief Executive (and his cabinet) and the civil service.

2.4.1 Hong Kong and the International Climate Change Regime

The United Nations Framework Convention on Climate Change (UNFCCC), supplemented by the Kyoto Protocol, underpins the formal treaty regime that has been created to coordinate inter-state action on climate change. While the UNFCCC contains provisions that encourage countries to undertake a broad range of actions to reduce their GHG emissions, the Kyoto Protocol creates legally binding emissions reduction targets for developed countries. These targets aggregate to an average 5 per cent emissions reduction compared to 1990 levels over the five-year period 2008 to 2012. Structured on the principle of common but differentiated responsibility enshrined in the UNFCCC, the Kyoto Protocol only imposes emissions reduction targets on developed countries.[34]

A few words on Hong Kong's unique status in international law are necessary to shed light on its involvement in the UNFCCC regime. Hong Kong was ceded by China to Britain under the terms of the 'Unequal Treaties' that were concluded to end the Opium Wars during the Qing Dynasty.[35] After 150 years of British rule, negotiations between the Chinese and the British led to the agreement that Hong Kong would be restored to the People's Republic of China.[36] On 1 July 1997, Hong Kong became a Special Administrative Region (SAR) of China. It should be noted that as a SAR, Hong Kong enjoys a high degree of autonomy. Under the Basic Law, the constitutional document of the HKSAR, which came into effect on 1 July 1997, '[t]he socialist system and policies shall not be practised in the Hong Kong Special Administrative Region, and the previous capitalist system and way of life shall remain unchanged for 50 years'.[37]

While China and the UK signed the UNFCCC in 1992 and the Kyoto Protocol in 1998, Hong Kong cited its 'unique status' during this period to stay out of the climate negotiations. As Zhao (2011) has pointed out, had Hong Kong joined the UNFCCC as a territory of the UK and hence obtained 'developed country' status as in the case of the Montreal Protocol, it would have had to fulfil the obligations of an Annex-I country (that is, it would have had to meet legally binding GHG emissions targets). These were obligations that Hong Kong was reluctant to undertake. Therefore, Hong Kong stayed outside the UNFCCC regime until China formally extended the UNFCCC and the Kyoto Protocol to the SAR in 2003. 'The practical benefits for Hong Kong to join the two treaties as part of China, a developing country, are too obvious' (Zhao 2011, p. 72). China, including Hong Kong, does not have to meet any GHG targets under international law and is merely required to formulate, report and regularly update national climate change programmes (Article 12 of the UNFCCC).

In summary, prior to 2003, the attitude of the Hong Kong government towards climate change was to evade and dodge legal responsibility as far as possible because climate change mitigation was perceived to be costly and would constrain economic growth. This perception of climate change (and environmental issues in general) continues to persist although the rhetoric is changing in response to the political demands of the local, national and international communities. The result has been a series of mitigation initiatives, mostly initiated by the government and pushed forward 'in a "muddling through" rather than coordinated fashion' by non-governmental stakeholders (Francesch-Huidobro 2012).

2.4.2 Institutional Inertia

In 2007, an Interdepartmental Working Group on Climate Change, comprising five bureaus and 16 departments, was established to coordinate the government's efforts in climate change mitigation and adaptation across agencies and departments.[38] However, this Working Group has yet to demonstrate its success in doing so – climate change continues to be tackled on a sectoral basis and within bureaucratic portfolios. Francesch-Huidobro's (2012) research shows that although the relevant policy bureaus and stakeholders demonstrate a good understanding of and familiarity with the challenges posed by climate change, knowledge management within and across the government remains hierarchical and sectorally defined. Such an approach hinders the development of a holistic approach towards climate change mitigation and adaptation (Francesch-Huidobro 2012).

Francesch-Huidobro argues that the Hong Kong government does not have the ability to formulate a coherent climate policy because the government does not enjoy the support and cooperation of the civil service. She attributes this situation to five main factors. First, following the Handover in

1997, there has been a shift from an executive-led (civil service elite-led) style of governance to a highly politicized non-civil service governance structure with politically appointed individuals heading policy bureaus. Second, the frequent interventions by the Chief Executive (head of the government) in setting very specific policy directions have undermined the leading role of the civil service in formulating policy and created a disincentive for civil service cooperation. Third, following the departure of Anson Chan, a powerful policy entrepreneur who was Chief Secretary for Administration until 2001, the Chief Secretary Committee lost a significant amount of its capability to broker policy coordination across government agencies. Fourth, there is a common perception amongst civil servants that the Chief Executive's priority of working together with the civil service to implement policy is secondary to that of following the dictates of the central government in Beijing and appealing to popular opinion to shore up the government's flagging popularity and lack of legitimacy (Cheung 2005, 2010; England 2012). Finally, the belief that under the new civil service regulations the Chief Executive has little power to remove civil servants has fostered a certain air of invincibility. This belief has also had the effect of reducing civil servants' incentive to cooperate with the Chief Executive. These factors collectively hamper the government's ability to put together and implement a coherent and institutionalized climate change adaptation policy.

2.5 CONCLUSION

The absence of governmental action has prompted civil society and corporations to be first movers on climate change. However, without a strong signal from the government such as laws and regulations to compel corporations to address climate change risks, the compliance-driven business sector in Hong Kong will not invest in adaptation measures (Shu and Schroeder 2010, p. 293). As for civil society, it is largely marginalized by the governmental machinery and there are few channels for civil society to contribute towards policy-making (Francesch-Huidobro 2007). There are therefore great limitations to how far private action can fill the climate governance gap in Hong Kong. As this chapter has shown, Hong Kong is highly vulnerable to climate change and there is a wealth of technical expertise and awareness about the risks and solutions. However, the lack of political will and institutional politics have prevented the government from formulating a coherent adaptation strategy and from working with community groups, non-governmental organizations and businesses. Therefore, despite its wealth and human talent, Hong Kong has a long way to go in meeting the challenges of climate change.

NOTES

1. The PRD also refers to the network of cities (Guangzhou, Shenzhen, Zhuhai, Dongguan, Zhongshan, Foshan, Huizhou, Jiangmen and Zhaoqing, and the Special Administrative Regions of Hong Kong and Macau) that collectively form a 'mega region' with an estimated population of 120 million people (Vidal 2010).
2. UN Habitat (2008, p. 79) points out that Hong Kong's Gini Coefficient of 0.53 makes it the most unequal city in all of Asia. For discussion, see Cheung (2011).
3. For historic rainstorm data and the impact of severe rainstorms, see HKSAR Government Drainage Services Department (2011).
4. This figure includes losses from landslides, flooding, flight disruptions, property damage and loss of business hours (Greenpeace East Asia 2009).
5. See, for example, European Commission (2010).
6. HKSAR Government Environmental Protection Department (2010), Appendix C 'Vulnerability and adaptation assessment', table 2.9, p. C-40.
7. Ibid, p. C-51.
8. Ibid, Appendix C, p. C-53.
9. Ibid, p. C-54.
10. Ibid, p. C-55.
11. Ibid, p. C-56.
12. HKSAR Government Environment Bureau (2010), p. 34.
13. Ibid, p. 43.
14. Ibid, p. 56.
15. Ibid, p. 34.
16. See Carbon Disclosure Project, which has pioneered a global system for the disclosure of corporate greenhouse gas emissions and climate change risk information, available at https://www.cdproject.net/en-US/Pages/HomePage.aspx (accessed 16 August 2012).
17. HKSAR Government Environmental Protection Department (2010), Appendix C 'Vulnerability and adaptation assessment', table 2.9, p. C-57.
18. Ibid.
19. See 'Responding to Diseases' in Section 2.3 below.
20. HKSAR Government Environmental Protection Department (2010), Appendix C 'Vulnerability and adaptation assessment', table 2.9, p. C-59.
21. London's Climate Change Adaptation Strategy (2010), available at http://www.london.gov.uk/climatechange/strategy (accessed 19 August 2012).
22. HKSAR Government Environmental Protection Department (2010), Appendix C 'Vulnerability and adaptation assessment', table 2.9, p. C-82.
23. For an interesting discussion of how ethics, knowledge, culture and risk perception shape a society's adaptive capacity, see Adger et al. (2009).
24. HKSAR Government Environmental Protection Department (2010), Appendix C 'Vulnerability and adaptation assessment', table 2.9, p. C-85.
25. Ibid.
26. Ibid, p. C-90.
27. The Hong Kong Observatory provides regular and detailed updates on weather conditions across Hong Kong and contains a wealth of educational resources, available at http://www.hko.gov.hk/contente.htm (accessed 21 August 2012).
28. HKSAR Government Security Bureau (2012), chapter 1, p. 2. For more information on the Daya Bay contingency plan, see HKSAR Government (2012b).
29. For a gripping account of the SARS outbreak, see Greenfeld (2006).
30. For example, on 3 December 2007, the health authorities of mainland China, Hong Kong and Macau carried out a joint exercise to test their cooperation and coordination in mounting an emergency response in the event of a cross-border incident of avian flu involving human cases. This was the second joint exercise organized under the Cooperation Agreement on Response Mechanism for Public Health Emergencies that the

three jurisdictions signed in 2005 (HKSAR Government Department of Health Centre for Health Protection 2007).

31. The Stanley Ho Centre for Emerging Infectious Diseases was established after 'the SARS outbreak in Hong Kong highlighted the lack of expertise in infectious diseases and hence effective measures in its control' (Stanley Ho Centre for Emerging Infectious Diseases 2012).
32. For discussion of the 2005 International Health Regulations, see Fidler and Gostin (2006).
33. HKSAR Government Environmental Protection Department (2010), Appendix C 'Vulnerability and adaptation assessment', table 2.9, p. C-93.
34. For discussion of the principle of common but differentiated responsibility, see Rajamani (2006) and Hey (2009).
35. For discussion of the Opium Wars, see Fay (1997) and Hanes and Sanello (2002).
36. The Joint Declaration of the Government of the United Kingdom of Great Britain and Northern Ireland and the Government of the People's Republic of China on the Question of Hong Kong was signed in Beijing on 19 December 1984. On 27 May 1985, instruments of ratification were exchanged and the agreement entered into force. It was registered at the United Nations on 12 June 1985. Article 1 of the Joint Declaration states '[t]he Government of the People's Republic of China declares that to recover the Hong Kong area (including Hong Kong Island, Kowloon and the New Territories, hereinafter referred to as Hong Kong) is the common aspiration of the entire Chinese people, and that it has decided to resume the exercise of sovereignty over Hong Kong with effect from 1 July 1997' and Article 2 states '[t]he Government of the United Kingdom declares that it will restore Hong Kong to the People's Republic of China with effect from 1 July 1997'.
37. Article 5 of the Basic Law of HKSAR.
38. For the Working Group's terms of reference and membership, see Legislative Council, Panel on Environmental Affairs (Subcommittee on Improving Air Quality), Meeting on 13 January 2009, Ref. CB1/PS/3/08.

REFERENCES

Abbott, C. (2012), 'Drought crop damage worsens, ethanol waiver urged', Reuters, 10 August, available at http://www.reuters.com/article/2012/08/10/us-drought-idUSBRE8781E320120810 (accessed 16 August 2012).

Adger, W.N., S. Dessai, M. Goulden et al. (2009), 'Are these social limits to adaptation to climate change?', *Climatic Change*, **93**, 335–54.

Basic Law of HKSAR, available at http://www.basiclaw.gov.hk/en/basiclawtext/index.html (accessed 15 April 2013).

Chan, E.Y.Y., W.B. Goggins, J.J. Kim, S. Griffiths and T.K.W. Ma (2011), 'Help-seeking behavior during elevated temperature in Chinese population', *Journal of Urban Health*, **88** (4), 637–50.

Cheung, A.B.L. (2005), 'Hong Kong's post-1997 institutional crisis: problems of governance and institutional incompatibility', *Journal of East Asian Studies*, **5** (1), 135–67.

Cheung, A.B.L. (2010), 'Political trajectory of Hong Kong as part of China', *International Public Management Review*, **11** (2), 38–63.

Cheung, A.B.L. (2011), 'A city of unhappiness', *Hong Kong Journal*, July, available at http://www.hkjournal.org/archive/2011_fall/1.htm (accessed 16 August 2012).

Cheung, A.B.L. and I. Scott (eds) (2003), *Governance and Public Sector Reform in Asia: Paradigm Shifts or Business as Usual?*, London: RoutledgeCurzon.

Chu, S.Y. and H. Schroeder (2010), 'Private governance of climate change in Hong Kong: an analysis of drivers and barriers to corporate action', *Asian Studies Review*, **34** (3), 287–308.

England, V. (2012), 'Hundreds of thousands protest as Hu Jintao visits Hong Kong', *Guardian*, 1 July, available at http://www.guardian.co.uk/world/2012/jul/01/protest-hu-jintao-hong-kong (accessed 24 August 2012).

Enright, M.J., E.E. Scott and Enright, Scott & Associates (2005), *The Greater Pearl River Delta*, 3rd edn, Report commissioned and published by Invest Hong Kong of HKSAR Government, available at http://www.investhk.gov.hk/doc/InvestHK_GPRD_Booklet_English571.pdf (accessed 10 January 2013).

Ensor, J. and R. Berger (2009), 'Community-based adaptation and culture in theory and practice', in W.N. Adger, I. Lorenzoni and K. O'Brien (eds), *Adapting to Climate Change: Thresholds, Values, Governance*, Cambridge: Cambridge University Press, 227–239.

European Commission (2010), *Climate Action: Adaptation to Climate Change*, available at http://ec.europa.eu/clima/sites/change/how_will_we_be_affected/sea_level_rise_en.htm (accessed 25 January 2013).

Fay, P.W. (1997), *The Opium War, 1840–1842: Barbarians in the Celestial Empire in the Early Part of the Nineteenth Century and the War by Which They Forced Her Gates*, 2nd edn, Chapel Hill, NC: University of North Carolina Press.

Fidler, D.P. and L.O. Gostin (2006), 'The new international health regulations: an historic development for international law and public health', *Journal of Law, Medicine and Ethics*, **34** (1), 85–94.

Francesch-Huidobro, M. (2007), 'Impact of government-NGO relations on sustainable air quality in Singapore and Hong Kong compared', *Journal of Comparative Policy Analysis*, **9** (4), 383–404.

Francesch-Huidobro, M. (2012), 'Institutional deficit and lack of legitimacy: the challenges of climate change governance in Hong Kong', *Environmental Politics*, **21** (5), 791–810.

Frumkin, H., J. Hess, G. Luber, J. Malilay and M. McGeehin (2008), 'Climate change: the public health response', *American Journal of Public Health*, **98** (3), 435–45.

Greenfeld, K.T. (2006), *China Syndrome*, London: Penguin Books.

Greenpeace East Asia (2009), *Black Rain: Hong Kong and Climate Change*, available at http://www.greenpeace.org/eastasia/news/stories/climate-energy/2009/hongkong-rainstorm/ (accessed 16 August 2012).

Hanes, W.T. and F. Sanello (2002), *The Opium Wars: The Addiction of One Empire and the Corruption of Another*, Naperville, IL: Source Books.

Hey, E. (2009), 'Common but differentiated responsibilities', in R. Wolfrum (ed.), *Max Planck Encyclopedia of Public International Law*, Oxford: Oxford University Press, available at http://opil.ouplaw.com/oxlaw/search?conr=Hey,%20Ellen (accessed 16 August 2012).

HKSAR Government (2007), *Hong Kong 2030 Planning Vision and Strategy*, available at http://www.pland.gov.hk/pland_en/p_study/comp_s/hk2030/eng/finalreport/ (accessed 16 August 2012).

HKSAR Government (2008), *Total Water Management in Hong Kong*, available at http://www.wsd.gov.hk/filemanager/en/share/pdf/TWM.pdf (accessed 17 August 2012), p. 16.

HKSAR Government (2012a), *Hong Kong: The Facts*, available at http://www.gov.hk/en/about/abouthk/facts.htm (accessed 16 August 2012).

HKSAR Government (2012b), *Daya Bay Contingency Plan*, available at http://www.dbcp.gov.hk/eng/info/index.htm (accessed 21 August 2012).

HKSAR Government Department of Health Centre for Health Protection (2007), 'Joint exercise on emergency response to infectious diseases', Press Release, 3 December, available at http://www.chp.gov.hk/en/content/116/10994.html (accessed 21 August 2012).

HKSAR Government Drainage Services Department (2011), *Historic Rainstorm Data*, available at http://www.dsd.gov.hk/EN/Flood_Prevention/Our_Flooding_Situation/index.html (accessed 16 August 2012).

HKSAR Government Environment Bureau (2010), *Hong Kong's Climate Change Strategy and Action Agenda: Consultation Document*, available at http://www.epd.gov.hk/epd/english/climate_change/files/Climate_Change_Bookle_E.pdf (accessed 14 April 2013).

HKSAR Government Environmental Protection Department (2010), *A Study of Climate Change in Hong Kong-Feasibility Study*, available at http://www.epd.gov.hk/epd/english/climate_change/consult.html#cc_study_report (accessed 16 August 2012).

HKSAR Government Planning Department (2011), *Hong Kong Planning Standards and Guidelines as at August 2011*, available at http://www.pland.gov.hk/pland_en/tech_doc/hkpsg/full/index.htm (accessed 21 August 2012).

HKSAR Government Security Bureau (2012), *HKSAR Emergency Response System*, available at http://www.sb.gov.hk/eng/emergency/ers/ers.htm (accessed 21 August 2012).

Kyoto Protocol to the United Nations Framework Convention on Climate Change 37 ILM 22 (1998), entered into force on 16 February 2005.

Lee, S.H. (2003), 'The SARS epidemic in Hong Kong: what lessons have we learned?', *Journal of the Royal Society of Medicine*, **96** (8), 374–8.

Leung, Y.K., M.C. Wu and K.H. Yeung (2006), 'Climate forecasting – what the temperature and rainfall in Hong Kong are going to be like In 100 years', Hong Kong Observatory, available at http://www.science.gov.hk/paper/HKO_YKLeung.pdf (accessed 16 August 2012).

Leung, Y.K., K.H. Yeung, E.W.L. Ginn and W.M. Leung (2004), 'Climate change in Hong Kong', Hong Kong Observatory Technical Note No. 107.

Meyer, G., J. Farchy and J. Blas (2012), 'US drought threatens food price surge', *Financial Times*, 10 August, available at http://www.ft.com/intl/cms/s/0/e37a491a-e2e1-11e1-a463-00144feab49a.html#axzz23owx3Q35 (accessed 16 August 2012).

O'Brien, K., S. Eriksen, L. Sygna and L.O. Naess (2006), 'Questioning complacency: climate change impacts, vulnerability and adaptation in Norway', *Ambio*, **35** (2), 50–6.

Pauw, P. and M. Francesch-Huidobro (2010), 'Hong Kong', in P. Dricke, J. Aerts and A. Molenaar (eds), *Connecting Delta Cities: Sharing Knowledge and Working on Adaptation to Climate Change*, Rotterdam, C40, available at http://www.rotterdamclimateinitiative.nl/.../CDC/CDC%20Boek%20II.pdf (accessed 16 August 2012), pp. 100–9.

Rajamani, L. (2006), *Differential Treatment in International Environmental Law*, Oxford: Oxford University Press.

Shu Yi Chu and H. Schroeder (2010), 'Private Governance of Climate Change in Hong Kong: An Analysis of Drivers and Barriers to Corporate Action', *Asian Studies Review*, **34** (3) 287–308.

Stanley Ho Centre for Emerging Infectious Diseases (2012), *About Us*, available at http://ceid.med.cuhk.edu.hk/about_s.html (accessed 21 August 2012).

The Joint Declaration of the Government of the United Kingdom of Great Britain and Northern Ireland and the Government of the People's Republic of China on the Question of Hong Kong, available at http://www.cmab.gov.hk/en/issues/joint.htm (accessed 22 August 2012).

Tracy, A., K. Trumbull and C. Loh (2006), *The Impacts of Climate Change in Hong Kong and the Pearl River Delta*, Hong Kong: Civic Exchange, p. 15.

UN Habitat (2008), *State of the World's Cities 2008/2009: Harmonious Cities*, London: Earthscan, p. 79.

United Nations Framework Convention on Climate Change 31 ILM 849 (1992), entered into force on 21 March 1994.

US Climate Change Science Program (2008), *Impacts of Climate Change and Variability on Transportation Systems and Infrastructure: Gulf Coast Study, Phase I*, Report by the US Climate Change Science Program and the Subcommittee on Global Change Research, M.J. Savonis, V.R. Burkett and J.R. Potter (eds), Department of Transportation, Washington, DC.

US Environmental Protection Agency (2012), *Heat Island Effect*, available at http://www.epa.gov/hiri/mitigation/pavements.htm (accessed 21 August 2012).

US Environmental Protection Agency (2013), *Coastal Areas Impacts and Adaptation*, available at http://www.epa.gov/climatechange/impacts-adaptation/coasts.html (accessed 25 January 2013).

Vidal, J. (2010), 'UN report: world's biggest cities merging into "mega-regions"', *Guardian*, 22 March, available at http://www.guardian.co.uk/world/2010/mar/22/un-cities-mega-regions (accessed 16 August 2012).

Welford, R. (2009), *Climate Change Challenges for Hong Kong: An Agenda for Adaptation*, Hong Kong: CSR Asia and Hong Kong University.

WMO (World Meteorological Organization) (2006), *Statement on Tropical Cyclones and Climate Change*, WMO International Workshop on Tropical Cyclones, IWTC-6, San Jose, Costa Rica, November, available at http://www.wmo.int/pages/prog/arep/tmrp/documents/iwtc_statement.pdf (accessed 16 August 2012).

Zhao, Y. (2011), 'Responding to the global challenge of climate change – Hong Kong and "one country, two systems"', *Carbon and Climate Law Review*, **5** (1), 70–81.

3. The evolution of environmental policies in South Korea in response to climate change

Ki-Ho Kim, Hye-Jin Jung and Chankook Kim

3.1 INTRODUCTION

3.1.1 Rapid Economic Development at the Sacrifice of the Environment

Throughout the period from 1960 to 1980, spurred by the enthusiastic implementation of its five-year plans for economic development, the South Korean economy grew rapidly and consistently with annual growth rates of over 7 per cent. Such economic development reached its climax as South Korea successfully hosted the 1986 Asian Games and 1988 Olympics. Unfortunately, under this intensive economic development, there was little concern about exploitation of the environment. However, in the early 1990s, the society at large was hit hard by important environmental accidents, triggering a change in citizens' awareness of the environment and creating a social consensus on its importance and value.

One of the representative incidents occurred in 1991 when phenol was released into the Nakdong River, one of the four main rivers of South Korea, thereby polluting drinking water. This accident was caused as phenol concentrate flowed into the intake station in the headbay of Taegu, contaminated tap water, and resulted in severe health damage including natural abortion. As a result, the chairman of the company that caused the accident resigned, and the then Minister and Vice-Minister of the Environment took responsibility and were replaced. The citizens of Taegu – the direct victims of this accident – started group protests, and soon afterwards a nationwide boycott of the company's products was initiated (Figure 3.1). This incident drew attention to the inadequacy of provisions for testing drinking water and the Act on Special Measures for the Punishment of Environmental Offences was enacted in 1991. The accident provided an opportunity to direct people's attention to the gravity of environmental problems.

Source: Kyunghyang Shinmun (1991), 24 March, p. 1. Reproduced with permission.

*Figure 3.1   Response of citizens and government officials to the phenol
leak incident: angry citizens boycott the products of
perpetrating company, Doosan Electronics*

3.1.2   Active Reform of Environmental Policies and their Effect

Such incidents raised civic awareness of the need for environmental
improvement, and the government began to actively improve environmental policies. As South Korea became a member of the Organisation for
Economic Co-operation and Development (OECD) in 1996, its people
started to perceive that their country should attain the status of a 'developed' country in the environmental field. They became more aware of the
need for a pleasant environment that enhances quality of life. Accordingly,
in order to meet the increased civic demands for higher environmental
quality, the government began to develop manageable environmental
standards in various fields, such as air quality, water quality and waste
disposal, and to formulate environmental policies at the level of those of
advanced countries.

Thanks to the reinforcement of environmental standards, there have
been improvements in the air quality of Seoul over the past decade,
with a reduction of particulate matter density (PM10) from $65\mu g/m^3$
to 40 $\mu g/m^3$. Water quality has also been enhanced, with biochem-
ical oxygen demand (BOD) density in the Han River having improved
from 1.6 mg/L to 1.4 mg/L. Furthermore, during the same period, the

amount of domestic waste has reduced by nearly half from 49,191 tons per day to 25,419 tons per day. This is due to the fact that as a result of institutional control and civic participation, the recycling volume has increased almost three times from 8972 tons per day to 24,588 tons per day.

3.2   CLIMATE CHANGE IN KOREA

3.2.1   Trends and Characteristics of South Korea's Greenhouse Gas Emissions

South Korea is the ninth largest emitter of greenhouse gases (GHGs) in the world. According to the latest statistics on national GHG emissions, total emissions amount to 600 million tons (Figure 3.2), and the major source of emissions is energy consumption. This is due to the structure of the industrial sector, which is energy intensive and highly dependent on oil (that is, industries such as steel, chemicals, cement and paper). Therefore, it is expected that GHG emissions will consistently increase in the next decade. Compared with other countries, South Korea consumes a lot of energy not only in the industrial sector but also by the entire society (Figure 3.3). Primary energy consumption and electricity consumption per capita in South Korea in 2010 reached 91.5 per cent and 87.1 per cent of OECD averages, respectively, while the gross domestic product (GDP) per capita was around 74.4 per cent of the average GDP per capita in OECD countries (GIR 2012a).

Due to such a rapid increase of GHG emissions, the impact of climate change is serious in South Korea. The pace of climate change on the Korean Peninsula exceeds that of the world average. During the last century, the precipitation in six major cities of South Korea increased by 19 per cent. The number of days of heavy rain, with over 80 mm per day, has more than doubled since the 1970s. Between 1964 and 2006, while the sea level around the coast of the Peninsula has risen by about 8 cm, the level around Jeju Island has risen by 22 cm, more than three times the world average, inundating some of the coastal paths. Should the sea level continue to rise at this rate, major coastal areas amounting to more than four times the area of Seoul would be inundated by 2100. Between 1968 and 2008, the surface water temperature surrounding the Peninsula rose by an average of 1.31°C, which is much higher than the world average of 0.5°C (Kwon 2009).

Moreover, extreme climate phenomena have started to occur. The number of tropical nights (nights with a minimum temperature of 25°C)

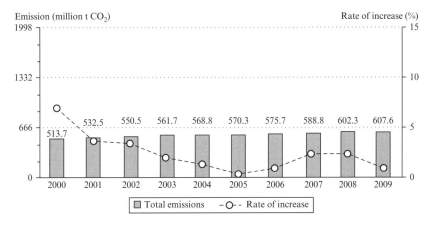

Source: Based on data from GIR (2012b).

Figure 3.2 *Total domestic GHG emissions and rate of increase*

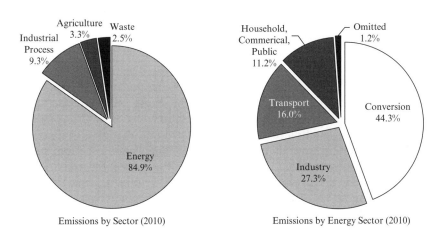

Emissions by Sector (2010)   Emissions by Energy Sector (2010)

Source: Based on data from GIR (2012b).

Figure 3.3 *Sectoral composition of GHG emissions*

in a year has increased from four to ten days over a century. Since 1973, for each decade, the duration of winter has decreased by 5.5 days. The strength of typhoons has also shown an increasing trend with the central pressure of typhoons decreasing by 7 hPa over the last decade. Due to the drastic increase in temperature, heavy rain and typhoons have occurred frequently, causing a large amount of casualties and property damage. For example, in 1998, localized heavy rains that started in the Jiri Mountain caused 324 casualties and property damage of around US$1 billion. In 1999, localized heavy rains in the northern part of Gyeonggi-do led to 64 casualties and around 25,000 refugees. In August 2002, the typhoon Rusa poured 870 mm of rain into the Gangneung area, setting the record for the maximum rainfall per day, and resulting in 246 casualties and property damage of more than US$4 billion, including the inundation of approximately 30,000 hectares of agricultural land. In 2003, another typhoon, Maemi, caused 130 casualties and property damage of approximately US$4.1 billion.

Climate change has also influenced people's lifestyles through changes in the ecosystem and human health. Since the 1990s, the frequency of high temperatures during the Korean summer has been generally increasing. In addition, during the period 1991–2000, the average death toll from July to August in Seoul was proportionate to the average temperature. During 1994–2005, 2127 people died due to the scorching heat, and, due to the temperature increase, malaria patients increased from five in 1994 to 22,227 in 2007.

Prospects of climate change in South Korea[1]
In December 2012, the Korean Meteorological Agency published the climate change forecast report for South Korea based on the Representative Concentration Pathways (RCPs) outlined for the 5th Intergovernmental Panel on Climate Change (IPCC) evaluation report. RCPs are the main GHG concentration pathway standards established by the IPCC as guidelines in its 5th evaluation report and the related information is updated to reflect the current trend of concentration change. Among the four main GHG concentration scenarios used in the RCPs, 2.6, 4.5, 6.0 and 8.5, the 2012 report for South Korea based its forecast on RCP 4.5 and RCP 8.5 scenarios (Table 3.1). RCP 4.5 is the hypothetical case equivalent to when a GHG reduction policy is successfully carried out; RCP 8.5 is the case when the current GHG emission trend is maintained; in other words, the business-as-usual scenario, which could be useful in analysing future policy making on climate change adaptation.

Annual mean temperature
In the RCP 8.5 scenario, South Korea's annual mean temperature is expected to increase by 1.4°C at the beginning of the twenty-first century, 3.2°C in the middle and 5.3°C at the end, starting from the current level of 12.5°C. Meanwhile, the RCP 4.5 scenario predicts less temperature increases: 1.2°C, 2.2°C and 2.8°C at the beginning, middle and end of the century, respectively. The RCP 4.5 scenario expects the annual mean

Table 3.1   Korea's prospects of climate change in the twenty-first century based on the RCP 4.5 scenario

	Current climate (1981–2010)	Beginning of the twenty-first century (2011–40)	Middle of the twenty-first century (2041–70)	End of the twenty-first century (2071–2100)
Mean temperature (°C)	12.5	13.7 (13.9)	14.7 (15.7)	15.3 (17.8)
Highest daily temperature (°C)	18.1	19.3 (19.5)	20.3 (21.2)	20.8 (23.4)
Lowest daily temperature (°C)	7.7	9.0 (9.1)	9.9 (11.0)	10.6 (13.1)
Precipitation (mm)	1307.7	1402.9 (1366.6)	1442.5 (1562.5)	1563.9 (1549.0)
Days of heat wave	10.1	11.7 (13.9)	15.3 (20.7)	17.9 (40.4)
Days of tropical night	3.8	6.1 (8.9)	14.8 (25.5)	22.1 (52.1)
Days of heavy rain	2.3	2.6 (2.3)	2.8 (3.3)	3.3 (3.2)

Note: Values in parentheses reflect changes based on the RCP 8.5 scenario.

Source: Based on data from the Korea Meteorological Administration (2012), pp. 70–95.

temperature of South Korea to be 15.3°C by the end of the twenty-first century, which is similar to the current temperature of one of South Korea's southernmost islands, Jeju.

Annual average precipitation
South Korea is expected to experience an increase in precipitation of 4.5 per cent at the beginning of the twenty-first century, 19.5 per cent in the middle and 18.5 per cent at the end compared to the current precipitation level, according to the RCP 8.5 scenario. This means that there will be a large-scale increase in the precipitation amount until the middle of the century, while the level is expected to be maintained towards the end of the century. The RCP 4.5 scenario also predicts Korea's annual average precipitation to increase up to 1563.9 mm by the end of the century, which is greater than the prediction from the RCP 8.5 scenario. The increasing movement of water vapor from the South relates to an overall increase of precipitation in this forecast.

Heat wave and tropical night
According to the RCP 8.5 scenario, the current average of 10.1 days of heat wave per annum will rapidly increase up to 13.9 days, 20.7 days and 40.4 days at the beginning, middle and end of the twenty-first century, respectively. The days of tropical night will also increase to 8.9 days, 25.5 days and 52.1 days for the same phases of the century, starting from the current level of 3.8 days per annum. The RCP 4.5 scenario predicts tropical nights will increase to 22.1 days by the end of the century; in effect, these predictions are far higher than the current average levels for South Korea.

Sea level rise
The RCP 8.5 scenario expects the sea level of both the Yellow Sea and the South Sea to rise by 65 cm and by 99 cm for the East Sea. The RCP 4.5 scenario predicts an increase of 53 cm for the sea level in the Yellow Sea and the South Sea and 74 cm for the East Sea. These levels in the Yellow Sea and the South Sea are slightly lower than the average rise in the global level, but the case of the East Sea is comparatively higher. The increase in heat transported by the Kuroshio current and the consequent rise in the temperature of the warm current are responsible for the conspicuous change of sea level in the East Sea.

The annual average wind speed, relative humidity and cloud amount are not expected to deviate much from the current climate in future.

3.3   CLIMATE CHANGE POLICIES IN SOUTH KOREA

The people of South Korea have given their critical attention to the issue of climate change since the mid 2000s. Currently, South Korea responds proactively to climate change and even considers it as a new means of growth.

3.3.1   International Efforts for Climate Change and the Preparation of the UNFCCC Comprehensive Counterplans

South Korea started participating in international efforts to prevent global warming with its signing the United Nations Framework Convention on Climate Change (UNFCCC) in December 1993. Under the Kyoto Protocol, South Korea was classified as a 'non-Annex I' country and had no obligations to reduce GHGs in the first commitment period (2008–12). However, in response to the UNFCCC, the Korean government established the UNFCCC National Organization in 1998 (in 2001, the name of the organization was changed to the UNFCCC Commission) to harmonize between economic development and climate change. In 2002, the Presidential Commission on Sustainable Development (PCSD) was established under presidential decree in order to actively respond to environmental problems of climate change and also to push forward the Korean society to a more sustainable one through policies ensuring participation of all sectors.

The aim of climate change policies made by the Korean government is to reduce the proportion of energy-intensive industries and increase that of less energy-consuming and high value-added industries such as information and communications and advanced technologies. By such measures, South Korea plans to join international efforts to reduce climate change by building an energy-saving economic structure and transforming the industrial sector structure. To implement such policies, the Korean government has prepared four UNFCCC counterplans since 1999. The Fourth UNFCCC Comprehensive Counterplan concentrated on five fields – international negotiation, GHG emission statistics, GHG reduction, climate change adaptation and technology development – to prepare for the post-Kyoto Mechanism[2] after the first commitment period ended in 2012. In order to make this happen, the Korean government aimed to invest approximately US$10 billion during 2009–13 for 'Low Carbon, Green Growth', of which the core indicator is the GHG reduction targets. The investment was expected to increase economic productivity and to grow larger when the emission trading system or the revenue from the carbon tax is introduced in the future. Quality of life is also expected to increase with the realization of high efficiency and environment-friendly buildings (Presidential Committee on Green Growth 2009).

3.3.2   Low Carbon, Green Growth and the National Strategy for Green Growth

In order to tackle the gravity of climate change on the Korean Peninsula, the former President Lee Myung-bak declared 'Low Carbon, Green Growth' as a national vision in 2008, and completed a comprehensive national roadmap to mitigate and adapt to climate change (Presidential Committee on Green Growth 2009). This plan was named 'National Strategy for Green Growth' and proposed ten policy directions for low carbon, green growth. According to this plan, the main contents regarding adaptation to climate change are classified as 'measures to adapt to climate change by field' and 'establishment of the base for adaptation to climate change'.

3.3.3   The National Climate Change Adaptation Strategies[3]

In accordance with the Framework Act on Low Carbon, Green Growth, the National Climate Change Adaptation Strategies (2011–15) were announced in 2010. These strategies are aimed at adaptation to climate change in a comprehensive and effective manner at the national level. They are composed of specific countermeasures to adapt to climate change in seven fields (public health, disaster reduction, agriculture, forestry, ocean/ fish industry, water resource management and ecosystem). The vision also includes development of a strong base for adaptation with measures in three areas (monitoring and forecast of climate change, adaptation in industry and energy, and education, public relations (PR) and international cooperation).The specific strategies are summarized in Figure 3.4.

3.3.4   Summary of Climate Change Mitigation and Adaptation Policies of South Korea

In 2001, the IPCC defined mitigation as 'an anthropogenic intervention to reduce the sources or enhance the sinks of greenhouse gases' and adaptation as 'adjustment in natural or human systems in response to actual or expected climatic stimuli or their effects, which moderates harm or exploits beneficial opportunities' (IPCC 2003, pp. 25, 11). Climate change policies, both mitigation and adaptation, in South Korea by sector are listed in Table 3.2.

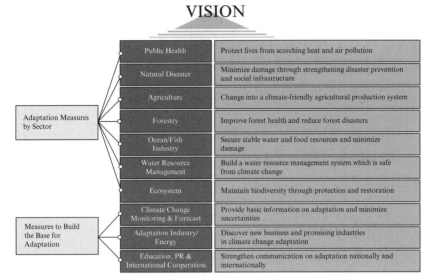

'Build a safe society and support green growth through climate change adaptation'

VISION

Public Health	Protect lives from scorching heat and air pollution
Natural Disaster	Minimize damage through strengthening disaster prevention and social infrastructure
Agriculture	Change into a climate-friendly agricultural production system
Forestry	Improve forest health and reduce forest disasters
Ocean/Fish Industry	Secure stable water and food resources and minimize damage
Water Resource Management	Build a water resource management system which is safe from climate change
Ecosystem	Maintain biodiversity through protection and restoration
Climate Change Monitoring & Forecast	Provide basic information on adaptation and minimize uncertainties
Adaptation Industry/ Energy	Discover new business and promising industries in climate change adaptation
Education, PR & International Cooperation	Strengthen communication on adaptation nationally and internationally

Adaptation Measures by Sector

Measures to Build the Base for Adaptation

Source: Based on data from MOE (2010), p. iv.

Figure 3.4   Vision and fields of national climate change adaptation strategies

3.4   IMPACT OF CLIMATE CHANGE IN URBAN AREAS

3.4.1   Main Cause of Climate Change in South Korea: Rise in Temperature due to Rapid Industrialization

In South Korea, disasters occurring due to climate anomalies are considered to be the most serious problems that need to be addressed through adaptation. In particular, heavy rainfall, scorching heat and heavy snowfall in urban areas are deemed to be the most severe. This is because cities are made up of artificial buildings and structures and thus are highly vulnerable to rapid changes of weather conditions. Furthermore, since primary assets are concentrated in the cities, they involve massive property damage and casualties. This characteristic of cities is demonstrated by statistics showing that damage estimates from typhoons and regional torrential rains have increased by 3.2 fold every decade (MOE n.d.).

South Korea has experienced drastic urbanization in recent decades.

Table 3.2 Comprehensive list of policies for responding to climate change in South Korea by sector

Policy sector		Strategy	Mitigation	Adaptation	Both	Year	Notes
National Level	Health	Evaluate health impact of scorching heat and build a system to monitor it.		O		2011	Law and legal framework have been enacted, and action plans for five years have been prepared by the national government.
	Agriculture	Develop crop varieties that adapt to climate change and new crops.		O		2011	Same as above.
	Forestry	Advance the system to prevent and reduce forest disasters.		O		2011	Same as above.
	Water Management	Build a system to predict floods.		O		2011	Same as above.
	Ecosystem	Build a system to evaluate vulnerability and a database on relevant information.		O		2011	Same as above.
	Monitoring and Forecast of Climate Change	Monitor climate change scenarios at the level of local government.		O		2011	In the process of being implemented.

Category	Measure				Year	Status
Transport	Reinforce the connection between public transportation and bicycles and encourage the development and dissemination of green cars.	O			2010	In the process of being implemented.
Reduction of Greenhouse Gases	Operate a target management system for GHG reduction and emission trading system (ETS).		O		2009	In the process of being implemented.
Green Technology and Industry	Promote green technology, green finance, green management, green company and green products.		O		2009	In the process of being implemented.
Green Construction	Make energy-efficient buildings compulsory.		O		2009	In the process of being implemented.
Energy and Resources Conservation	Disseminate renewable energy and smart grids.		O		2008	In the process of being implemented.
Municipal Level Transport	– Congestion Fee – Bus Rapid Transit System (BRT)	O			1996 2004	The congestion fee is applied to vehicles passing through Namsan tunnel, a gateway to downtown Seoul.
Air Quality	– Compressed Natural Gas (CNG) Bus Project – Special Act on the Improvement of Metropolitan Air Quality			O	2002 2003	

Table 3.2 (continued)

Policy sector	Strategy	Mitigation	Adaptation	Both	Year	Notes
Urban Park	– World Cup Park (Nanjido landfill site) – Seoul Forest Park		O		2002 2005	Seoul Forest Park was downzoned for use as parks on the site that was originally zoned for commercial and mixed use development.
Residential Development	Provide public open spaces and parks through the creation of greenways throughout the city (along the waterfront as well as within new residential developments).		O		2006	To be part of the 'Public Park Movement' for 10 million *pyeong* (approximately 8200 acres) of green space in Seoul (one *pyeong* for each person of the ten million population).
Education	Green Leadership Programme (established for undergraduate students at university and certified by the Minister of the Environment).		O		2011	The programme was established first at Seoul National University and is expected to spread to other universities throughout the country.

The rate of urbanization was only 40 per cent in 1960 but surpassed 90 per cent in the 2000s.[4] Meteorological research centres estimate that 20–30 per cent of South Korea's temperature increase during the twentieth century could be attributed to the urbanization effect. This estimation is based on the results of a study on the influence of urbanization on the rise in temperature, controlling for the effect of population size. According to the study on the heat island effect in Korean cities since the late 1960s, the heat island effect accounted for around 30–40 per cent of the rate of increase in temperature, and localized urbanization amplified not only the average annual temperature but also the average annual rate of change of maximum and minimum temperature (Koo et al. 2007). In the cases of Seoul, Gangneung, Taegu and Busan, which have sufficient data observed over a long period of time, industrialization and urbanization have proceeded at a fast pace since the late 1960s, and they show a distinct heat island effect. During 1974–97, an analysis of the average annual temperature of nine observation points in South Korea showed that the temperature rose by 0.96°C during this period, with a 1.5°C rise in urban areas and a 0.58°C rise in non-urban areas and coastal observatories.

3.4.2 Urbanization of Seoul and its Vulnerability to Climate Change[5]

In the case of Seoul, the capital of South Korea, due to high-density urbanization and the subsequent heat island effect, the average annual temperature has risen by 2.4°C over the last century, and the average annual rainfall has been increasing rapidly over the last decade. In particular, the number of days hit by heavy rain has increased four times in this period. In Seoul in 2011, there was heavy rain measuring 588 mm over three days, the highest rainfall ever recorded in South Korea. On an hourly basis, the precipitation in 1980 was around 95 mm per hour, but in 2009, it increased to around 110 mm per hour. Heavy rain pouring at 100–200 mm a day has become more frequent and the frequency of rain falling more than 50 mm or 80 mm per hour has also been increasing.

In July 2012, the Seoul Institute pointed out that the reason Seoul had become more vulnerable to natural disasters was the increase in precipitation due to climate anomalies and rapid urbanization. As a result of rapid urbanization, rainwater cannot be absorbed into the ground surface, that is, the impervious area and total area of buildings increased, causing the flow of rainwater to increase and aggravate flood damage. In 1994, the impervious area in Seoul was 302.8 km^2, covering 50.1 per cent of the total area. However, in 2008, it grew to 319.7 km^2, which is 53 per cent of the total area. Furthermore, the total area of buildings was 551.52 million m^2, a 1.6 fold growth compared to 2000. The expansion of the impermeable area, which increases the flow of rainwater on the land surface, due to the highly urbanized local environment is the major cause aggravating flood damage.

3.5 STRATEGIES FOR CLIMATE CHANGE ADAPTATION IN URBAN AREAS

3.5.1 Climate Change Adaptation Policies to Respond to Rapid Urbanization

In the C40 Seoul Summit for climate change in 2009, Ken Livingstone, former Mayor of London, emphasized the role of the city in responding to climate change. The fact sheet published by the C40 Cities Climate Leadership Group (n.d.) notes that the 'cities consume over two-thirds of the energy and account for more than 70 per cent of global CO_2 emissions, the most prevalent of the greenhouse gases (GHG)', while the cities are still growing in size, which, in other words, directs attention to the point that the '[c]ities have a responsibility to create solutions to climate change' (p. 1). Livingston, in line with these findings, also emphasized the significance of response at the municipal level rather than at the national level in humanity's battle against climate change.

In the case of Seoul, the concentration of the population, given rapid economic development and subsequent urbanization, heightens the threat of climate change. In order to respond to the threat, the Korean government is designing new policies and ways of implementation to boost adaptive capacity in the fields of health, disasters, forestry, water management, education and PR. These actions for adaptation are focused on making climate change policies that allow citizens to enjoy a safe and healthy city life.

This chapter introduces two major adaptation policies – the expansion of urban parks and the establishment of information technology (IT)-based climate change forecast and monitoring systems – which are representative for South Korea's metropolitan area, where 50 per cent of its population is concentrated.

3.5.2 Urban Parks for Cities

Two-thirds of the national territory of South Korea is mountainous. Ironically, however, there is a great lack of residential parks in the cities. Seoul, one of the global cities, provides merely around 5 m² of residential park for each person, which is much smaller than the 9 m² recommended

Source: Lee (2012), p. 14. Reproduced with permission.

Figure 3.5   School Forest project. The School Forest is a movement to green school playgrounds through cooperation between civil society and the government

by the Food and Agriculture Organization (FAO). This is because, compared to the rapid increase in housing demand in the past decades, the land available for development has been insufficient, and, as a result, the city has prioritized development to the supply of housing. Such development produces many side effects. The most representative side effect is the lack of infrastructure, including public open spaces and community facilities, to facilitate the city's functions.

As the value of parks rose, together with the residents' desire for quality of life, efforts to increase the downtown park area as a new form of city infrastructure have been initiated. Urban parks provide places to relax, promote urban aesthetics and generate a positive effect on the urban climate through mitigating sunshine and radiant heat and thus alleviating the heat island phenomenon. Furthermore, in case of a natural disaster, they could act as a refuge and fend off disasters by preventing flood damage and landslides.

Due to these advantages of urban parks, civil society and the government of Seoul have collaborated to initiate the '10 million *pyeong*[6] movement' through which the Seoul Forest, World Cup Park in Nanjido, city forests (School Forests) (Figure 3.5), green spaces on old public facility sites and new greenways have been built.

Generally, these parks were created on land where a large facility that originally existed was no longer needed, closed or moved to another area. Nonetheless, since there is a limit to park expansion by the public sector, policies have been made to require private developers to create a certain area of parks in their development. Representative among them are policies enforcing the building of a linear park through greenways (Figures 3.6

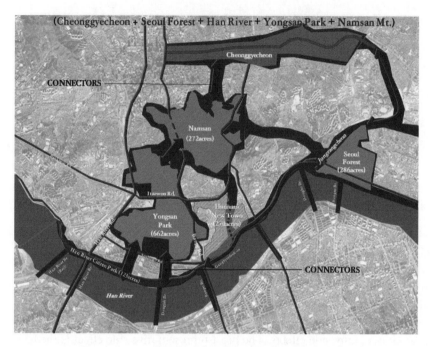

*Figure 3.6   Long-term vision for greenways in Seoul to network
through greenways (Cheonggyecheon → Seoul Forest →
Han River → Yongsan Park → Namsan) proposed by the Urban
Design Center*

and 3.7) and open spaces or pocket parks while implementing large-scale
development projects. Moreover, projects such as greening school playgrounds and the area along the river have been implemented to make parks
more accessible for the residents as well as the general public.

In order to design greenways in cities, it is necessary to set the criteria
and the directions of urban redesign in the process of urban redevelopment. In particular, the criteria to design and build greenways in urban
development projects are needed for the private sector. Based on urban
design works in past years, Kim and Moon (2006, (pp. 179–209) have proposed the following ten principles as guidelines for the creation of a more
climate-positive city.

1.   At city-wide scale: park area per person 10 m²~17 m².
2.   At project scale: 30~50 per cent of the total site for public open space.

Source: Kim (2009).

Figure 3.7   *Design proposal by the Urban Design Center for the creation of a greenway in a redevelopment project along the Han River*

3. Each residential unit within 250 m from the greenway.
4. Design green space first and build other elements around it.
5. Mixed use: residential, commercial and retail, educational and public open spaces.
6. Create public open spaces on private assets through the redevelopment process.
7. Create a long-term master plan alongside short-term applications.
8. Employ an integrative approach between multi-sectors as well as multi-departments.
9. Create a new governance: engage independent, private not-for-profit corporations.
10. Citizen participation: advocates for public cause.

In the process of making alternative urban designs, based on these rules and previous experiences, the following issues have been raised as paramount. The first relates to, in light of easy and convenient pedestrian accessibility, where to locate public spaces where people can relax and enjoy the company of others. Second, there is the issue of how to locate high and low buildings in order to make the space between the buildings more public and more pleasant by providing enough natural light into the courtyard. In this way, the dwelling units of the surrounding buildings can have enough daylight so that residents use less energy and electricity for lighting and heating. The third issue concerns how to reduce the use of private vehicles and thus make an environment where using public transportation and walking are more convenient.

Once the most fundamental and crucial elements such as those above, which have too easily been neglected in practice, are taken into consideration in urban redesign, they will help change people's lifestyles and create a city that can adapt to climate change.

3.5.3 IT-based Climate Change Forecast and Monitoring System[7]

Intense urbanization is not necessarily disadvantageous for climate change adaptation. As a matter of fact, South Korea is promoting ways to benefit from urbanization by making better use of the concentrated infrastructure and strengthening the IT-based climate change forecast system. As widely known, South Korea is a globally acknowledged IT power. More than 80 per cent of the households residing in South Korea have Internet access, and mobile broadband services are available across the entire territory. Taking advantage of these strengths, the government is reinforcing monitoring and forecast information services concerning climate change and vigorously implementing projects that utilize them.

As previously mentioned, South Korea primarily showed interest in adaptation through the National Plan for Climate Change Adaptation in 2008, and formulated the National Climate Change Adaptation Strategies (2011–15). The strategies are composed of specific countermeasures that focus on seven fields (public health, disaster reduction, agriculture, forestry, ocean/fish industry, water resource management and ecosystem) and measures to build the base for adaptation in three areas (monitoring and forecast of climate change, adaptation in industry and energy, and education, PR and international cooperation).

First among the three base measures, a monitoring and forecast system has been built based on the advanced IT infrastructure. The IT-based forecast and notification services based on this system are being relatively actively used in the fields of public health, disaster relief, agriculture and forestry (Table 3.3). For instance, using the basic data of local ultraviolet (UV) light indices, a UV light monitoring system has been built, and its contents have been connected with the Geographic Information System

Table 3.3 IT-based systems for adaptation by category

Category	System	Purpose	Content
Public Health	Ultraviolet (UV) light monitoring system	Due to excessive UV light exposure, there were approximately 60,000 premature deaths in 2000. Thus, the Korea Meteorological Administration forecasts the UV light index of each region and provides directions for each level of UV light exposure.	Provides UV light strength and directions according to five levels of UV light exposure for each region.
Disaster	National Disaster Safety Center application	In order to reduce the flood damage that occurs each year, it is necessary to respond swiftly in case of a real occurrence. Thus, the Ministry of Public Administration and Security (MOPAS) and the National Emergency Management Agency (NEMA) developed the 'National Disaster Safety Center' application to enable people to safely cope with disasters.	Provides real-time text messages, codes of conduct by type of disaster, emergency first aid method etc.
Agriculture	Bioclimate information service	In the agricultural sector that is vulnerable to climate change, there is a variety of forms of damage. Thus, in order for the mandarin farms to respond to climate change, Jeju Island provides the 'bioclimate information service'.	Provides the infection risk of pests (common scab, black speck and scale) and time series data on time courses.
Forestry	Forest fire on-site response system	Due to the increase of dry days and combustibles in forests, large-size forest fires are frequent. Thus, in order to protect the residents and facilities, the Korea Forest Research Institute of the Korea Forest Service developed and provides the 'forest fire on-site response system' based on tablet PCs.	Provides tactics to extinguish the fire and countermeasures to protect residents and facilities, forecasts fire expansion, calculates the area of damage and locates the facilities vulnerable to the forest fire.

Table 3.3 (continued)

Category	System	Purpose	Content
Ocean/Fish Industry	Voice broadcast of ocean weather	Weather changes on sea are directly linked with the lives of fishermen. Therefore, this system swiftly provides information on the ocean conditions through media such as voice broadcast, short message service (SMS), multimedia messaging service (MMS) and Twitter for workers in ocean-related industries.	Provides special weather reports, ocean weather forecasts (daytime), live coverage of weather for the fishing industry etc.
Weather Forecast	Mobile web for weather information	The rapidly changing abnormal weather and impact of climate change fluctuates with time and space. Thus, through the mobile web for weather information, this system provides fast and easy access to main weather information using smartphones.	Provides basic information such as current weather in the area and its neighbourhood, dangerous weather information (typhoon, earthquake and yellow dust), leisure weather etc.
Climate Change Forecast	National standard climate change scenarios	In order to evaluate the impact of climate change and produce countermeasures for adaptation, it is necessary to have information on the future scenarios of the trend and variability of climate change and extreme climates. Thus, to support the national or local governments to make policies for climate change adaptation, South Korea provides detailed climate change scenarios (1km resolution) for the country.	Calculates the average, maximum and minimum temperatures and precipitation, reflecting the altitude of each region and slope and direction of mountains.

(GIS) via the Internet to offer the information to the people. Especially for disasters, such as meteorological disasters and forest fires that require swift responses, smartphone applications have been developed to provide information on national disasters and climate disasters and codes of conduct in case of emergencies.

Table 3.4 shows the basic national data, tool and media used to build the information system by category. Table 3.5 gives detailed information on one specific example of information systems – the forest fire on-site response system – that has interesting characteristics.

3.6   CONCLUSION

Over the last four decades, South Korea has achieved rapid economic development at the expense of the environment and has become the 13th largest economy with a trade volume of over US$1 trillion. This was possible because economic development was set as the top priority of the country. Apart from the economy, however, the environmental situation of South Korea continued to deteriorate. In 2001, it was ranked 136th among 142 countries in the world by the Environmental Sustainability Index (ESI) at the Davos Forum. At that time, the South Korean people were shocked and realized that one of the adverse outcomes of South Korea's economic development was the rapid deterioration of the quality of its environment.

Over the past decade, there have been tremendous changes in the environmental policies in South Korea. As a result, in the announcement of the EPI in the Davos Forum in 2012, South Korea's experience was selected as one of the best cases. In this chapter, we have attempted to share our trials and errors in the past and small but meaningful successes over the last decade with people from other countries.

Diverse societal sectors, with civil society at the centre, at both the national and municipal levels have collaborated to improve the environmental situation in South Korea. There are several landmark policies of high significance. First, at the national level, the Special Act on the Improvement of Metropolitan Air Quality enacted in 2003 vastly improved the air quality in the Seoul Metropolitan Area, which includes Seoul, Gyeonggi and Incheon and provides home for half of the population. The Act became a catalyst and, at the municipal level, triggered the government of Seoul to establish the Master Plan for Management of Air Quality of the Seoul Metropolitan Area to improve the environment of Seoul, to the extent that people can drive in their cars with the windows open in downtown Seoul. Under this Master Plan, the 10 million *pyeong* movement was initiated and the Nanjido landfill site was transformed into

Table 3.4 Data, tool, media and users of each system by category

Category	System	Raw data	Tool	Media	User
Public Health	UV light monitoring system	UV light indices by region.	Geographic Information System (GIS)	Internet	All people.
Disaster	National Disaster Safety Center application	Information on special weather reports, national codes of conduct and information on facilities nearby	Application	Smartphone	All people (smartphone users).
Agriculture	Bioclimate information service	Infection risk of pests (common scab, black speck and scale), route and time, precipitation, information on the season that pests occur etc.	GIS	Internet	Workers in mandarin production in Jeju area.
Forestry	Forest fire on-site response system	Pictures and videos including the coordinates information of where the forest fire occurred, information on tactics to extinguish the fire etc.	Application	Tablet PC	All people (PC users).
Ocean/Fish Industry	Voice broadcast of ocean weather	Ocean forecasts by sea, information on typhoons and tsunamis, live coverage of light beacon weather, live coverage of weather for fish industries, information on live weather in marine zones etc.	Application	Smartphone	Workers in ocean-related industries (smartphone users).
Weather Forecast	Mobile web for weather information	Nationwide weather videos, information on dangerous weather (typhoon, earthquake and yellow dust), leisure weather (sea, mountain and daily life) etc.	Application	Smartphone	All people (smartphone users).
Climate Change Forecast	National standard climate change scenarios	Forecasts on the average, maximum and minimum temperatures and precipitation of each region (within 1 km) over the next century.	GIS	Internet	All people.

Table 3.5   Information on the forest fire on-site response system

System	Forest fire on-site response system
Main Contents	This system delivers pictures and videos including the coordinated information to the forest fire situation room and allows for the swift formulation of countermeasures to protect residents and facilities. Furthermore, it is possible to forecast the expansion of the fire and calculate the area of damage right away on the site and respond after locating the facilities nearby that are vulnerable to the fire.
Example and Application	 
	The example shows the screen capture of the system being operated when there was a forest fire in a region in Gwanak-gu, Seoul, on 22 November 2011. Once you enter the location of where the fire broke out and weather information as in the left picture, it is possible to get a live forecast of the fire expansion as shown in the right picture.

the World Cup Park. Furthermore, a high-density mixed use development site was downzoned into the Seoul Forest. By building the Bus Rapid Transit (BRT) system, with bus lanes along the centre of the road, the city government improved the public transportation service so that citizens would reduce their use of private vehicles.

In the early 2000s, in practical terms, there was no clear division between the two concepts of adaptation and mitigation in responding to climate change in South Korea. Many policies had double aspects – of both adaptation and mitigation. It was after the announcement of the Low Carbon, Green Growth policies by the former President Lee Myung-bak along with the enactment of the Framework Act and National Strategies that South

Korea began to make a practical distinction between adaptation and mitigation in responding to climate change.

In order to minimize the impact of climate change in urban areas, public policies have been implemented to create large public parks in the downtown area, green school playgrounds and linear parks through greenways, as well as requiring private developers to secure a minimum area of parks in their development projects. Furthermore, IT-based forecast services for climate change adaptation have been strengthened to promptly provide a variety of climate change-related information to people in the fields of public health, disaster reduction, agriculture and forestry, and consequently protect people and property from various climate disasters. Adaptation is focused on such policies since, due to rapid urbanization, most of the South Korean population is concentrated in cities. The most effective adaptation policies are considered to be a reduction of the danger of disasters by developing the city's capacity to respond to natural disasters and the prompt delivery of information.

As previously discussed, although cities comprise only 2 per cent of the Earth's surface, 80 per cent of the world's GHG emissions are produced by cities. Moreover, it should be remembered that in the war against climate change, victory depends on the policies of the local or city government. Therefore, the challenge we face, among others, is to change the lifestyle of the urbanites from an energy-guzzling one today to an energy-saving one. This is possible by enhancing people's awareness of climate change adaptation and city infrastructure and, furthermore, actively transforming the methods of urban planning, local planning, housing plans and architectural planning. A new paradigm of the policies related to cities, architecture and parks awaits us.

ACKNOWLEDGEMENTS

Special thanks are due to Mr Kook-Hyun Moon. Ever since Korea ranked 136th in the ESI ranking at Davos Forum, Mr Moon, as a civic leader and advisor, had tried to help the Korean government both at the national and municipal level change the environmental policies. He coordinated and advised the creation of mitigation and adaptation policies in the face of climate change and challenges to sustainability under the administration of former President Moo-Hyun Roh as well as under the municipal administration of former Mayor of Seoul, Myung-Bak Lee. As the living witness of and key actor contributing to the evolutionary change of Korea's environmental policies, Mr Moon provided the authors of this chapter with plenty of sources and information that helped them easily make sense of the history of Korean environmental policy. The authors regret that Mr Moon could not participate as one of the co-authors of this chapter and are grateful for his contribution.

Gratitude is also due to Ms Yeonwoo Lee and Ms Ji Yeong Yoo for their professional editorial service. In order to capture the essence of the cases of South Korea, the first drafts of the chapter were written in the Korean language. Ms Lee and Ms Yoo then helped the authors to translate the draft into English.

NOTES

1. This section draws on Korea Meteorological Administration (2012), pp. 70–95.
2. In the Doha Climate Change Conference (COP18), the Kyoto Mechanism has been extended to 2020.
3. This section draws from MOE (2010), p. 6.
4. Urbanization rate refers to the proportion of the total population living in urban areas.
5. This section draws on Department of Climate and Atmosphere (2010), pp. 7–8.
6. A *pyeong* is a Korean metric amounting to around 3.3 m^2.
7. This section draws on MOE (2011).

REFERENCES

C40 Cities Climate Leadership Group (n.d.), *Fact Sheet: Why Cities?*, available at http://www.c40cities.org/media (accessed 8 February 2013).

Department of Climate and Atmosphere (2010), *Climate Change Adaptation: The Path We All Have to Take Together*, Seoul: Seoul Metropolitan Government.

GIR (Greenhouse Gas Inventory and Research Center of Korea) (2012a), *Greenhouse Gas Statistics*, available at http://www.gir.go.kr/og/hm/gs/a/OGHMGSA010.do (accessed 1 May 2013).

GIR (Greenhouse Gas Inventory and Research Center of Korea) (2012b), *National Statistics on Greenhouse Gas Emissions*, available at http://www.index.go.kr/egams/stts/jsp/potal/stts/PO_STTS_IdxMain.jsp?idx_cd=1464&bbs=INDX_001 (accessed 31 January 2013).

IPCC (Intergovernmental Panel on Climate Change) (2003), 'Glossary of terms', in *Climate Change 2001 Synthesis Report*, Annex B, available at http://grida.no/climate/ipcc_tar/vol4/english/pdf/annex.pdf (accessed 8 February 2013).

Kim, K.-H. (2009), 'Design for sustainable environment: city design initiatives through greenways', in *Design for Social Innovation*, IDCC 2009 Proceedings, Seoul, pp. 48–71.

Kim, K.-H. and K.-H. Moon (2006), *The Life of the City: Greenways*, Seoul: Random House.

Koo, K.-S., G.-E. Boo and W.-T. Kwon (2007), 'An analysis of the urbanization effect in temperature change in South Korea using the maximum and minimum temperature', Korea Meterological Society, *Atmsophere*, **17** (2).

Korea Meteorological Administration (2012), 'Forecast of future climate change', in *Prospects of Korea's Climate Change*, Seoul: Korea Meteorological Administration, Chapter 4.

Kwon, W.-T. (2009), *Understanding Climate Change on the Korean Peninsula*, Seoul: National Institute of Meteorological Research.

Kyunghyang Shinmun (1991), 'Anger is aflame', Kyunghyang Shinmun, 24 March.

Lee, K.-O. (2012), *Green Dream, Green City working together; The Story of Seoul Green Trust*, Seoul: Seoul Green Trust.

MOE (Ministry of the Environment) (2010), *National Climate Change Adaptation Strategies Based on the Framework Act on Low Carbon, Green Growth (2011–2015)*, Gwacheon, Republic of Korea: Ministry of the Environment.

MOE (Ministry of the Environment) (2011), *Environmental Statistics Yearbook*, Gwacheon, Republic of Korea: Ministry of the Environment.

MOE (Ministry of the Environment) (n.d.), *Climate Change and Disaster: Impact and Prospects from 2011~2015 National Climate Change Response*, Gwacheon, Republic of Korea: Ministry of the Environment.

Presidential Committee on Green Growth (2009), *National Strategy for Green Growth*, available at http://www.greengrowth.go.kr/?page_id=31 (accessed 1 February 2013).

4. Climate-proofing a concrete island: improving state and societal climate adaptation capacities in Singapore

Sofiah Jamil

Yesterday's event, I was told by the PUB [Public Utilities Board], occurs once every 50 years. It could be tomorrow. But we have to plan accordingly. Most importantly, we have to have a proper drainage system . . . [that is] being continually upgraded, and a proper response system. (Dr Yaacob Ibrahim, Singapore's Minister for the Environment and Water Resources, 20 November 2009, quoted in Popatlal 2009)

The quote above refers to Dr Yaacob's response on flash floods that had occurred in various parts of Singapore on 19 November 2009. While flash floods in Singapore are considered to be a low-risk occasional occurrence particularly during the monsoon season, there has been a growing visibility of higher flood levels in recent years. Approximately half a year after this statement about intense flash floods being a once in 50 years phenomenon, however, Singapore again experienced flash floods, but this time in the heart of the city-state's prime commercial area, Orchard Road in June 2010 and then again in June 2011. Images of the flooded busy traffic junctions were a stark contrast to the glitzy buildings surrounding it and left many members of the public shocked that such an event, commonly seen in Jakarta or Manila, was happening in relatively disaster-free Singapore. While it is clear that there has been a higher level of rainfall in Singapore in recent years, the repeated flash floods have raised questions over Singapore's readiness to adapt to changing weather conditions and the extent to which Singapore's track record of sustainable urban planning policies can address these concerns.

The Orchard Road floods also demonstrated that despite Singapore's achievements as a beacon of sustainable development thus far, the city-state of 714.3 square kilometres (DOS Singapore 2012) is not immune to the effects of climate change. On the one hand, Singapore's pragmatic economic development and urban planning policies have allowed it to overcome socio-economic vulnerabilities and the dearth of natural resources since it gained independence in 1965. On the other hand, Singapore's massive rate of urbanization has the potential to exacerbate the direct impacts of climate change. Moreover, Singapore may suffer indirect impacts of climate change in the form of food insecurity. This is because Singapore is highly dependent on the global market for resources and food imports that may be at risk to weather-related events if effective mitigating measures are not taken. As such, if Singapore is to adapt to climate change, it must take into account both domestic and global impacts of climate change, which are compounded by the implications of urbanization.

In light of these developments, this chapter argues that while Singapore does possess advanced top-down technical and infrastructural capabilities for climate change adaptation (CCA) measures, the degree of bottom-up adaptive capacities and consciousness on climate change needed to complement government efforts are less pronounced. Section 4.1 highlights the direct and indirect impacts of climate change on Singapore. Section 4.2 focuses on government strategies to address the direct impacts of climate change, and highlights their weaknesses and how they can be improved. Section 4.3 examines efforts to address food security and how these efforts can be enhanced. What is clearly concluded from these sections (Section 4.4) is the fact that despite all the governmental efforts, there remains room for improvement, which can be spearheaded by Singapore's civil society.

4.1   IMPACT OF CLIMATE CHANGE ON SINGAPORE

Studies on climate change have shown how it has the potential to exacerbate existing environmental challenges. While Singapore has enjoyed relative protection from natural disasters in contrast to its neighbours – that is, not situated on the Pacific Ring of Fire (countries situated around the Pacific Basin and highly vulnerable to earthquakes) and protected from typhoons by Malaysia and Indonesia – it does not mean that the city-state is totally safe from these hazards, directly or indirectly. This section examines these direct and indirect impacts.

4.1.1   Direct Impacts

While the Intergovernmental Panel on Climate Change's (IPCC) estimates on global warming and sea level rise have been noted in previous chapters, it is important to take note of projections specific to Southeast Asia. Global warming projections for Southeast Asia are said to be fairly similar to the global mean warming – with global temperatures projected to rise by 1.1°C to 6.4°C and the predicted annual rainfall changes for Southeast Asia to range from –2 per cent to +15 per cent with a median change of +7 per cent (MEWR 2009). Moreover, the direct implications of increased temperatures and higher rainfall in Singapore are amplified by the processes of urbanization.

The first and perhaps most visible impact of climate change is flooding. Rising sea levels can cause flooding and ultimately a loss of coastal areas if little is being done to address the situation. This is particularly so given the fact that a substantial portion of Singapore's natural coastline has been altered via land reclamation to make room for industrial and trading zones. Flooding of these areas would cause substantial impediments in sustaining Singapore's economy. Flooding is also exacerbated by high levels of rainfall. In the case of the Orchard Road flood in 2010, aspects of urbanization have been recognized as contributing factors to the flooding, such as increasing the storm water run-off (Today Online 2012).

Second, climate change can have an adverse impact on public health, namely via increasing temperatures and higher rainfall, which can catalyse the spread of communicable diseases. Several studies (for example, Khasnis and Nettleman 2005; Patz and Olson 2006) have demonstrated how the higher temperatures have contributed to the spread of communicable diseases, such as malaria and dengue fever. With regard to dengue fever – of which Singapore has an average of 5000 reported cases per year (Grosse 2012) – higher rainfall also contributes to the spread of waterborne diseases by increasing the likelihood of breeding grounds for the Aedes mosquito.

Third, rising temperatures in Singapore will be further exacerbated by the Urban Heat Island (UHI) phenomenon, which is the condition when an urban area has higher temperatures than its surrounding rural areas. As seen from Figure 4.1, there is a difference of up to 4.01°C between the temperatures of well-planted areas and highly urbanized areas (such as the Central Business District, CBD) in Singapore (Wong 2011). In addition, a study conducted on temperatures of areas in and around Bukit Batok Nature Park (BBNP) and neighbouring public housing apartments (commonly known as Housing and Development Board (HDB) flats) (Wong 2011) revealed that urban areas in closer proximity with greenery would still have a lower temperature than urban areas located further away from greenery (Figure 4.2). The higher temperatures in urban areas would likely translate to an increasing use of energy (such as for air-conditioning). Given the fact that the bulk of Singapore's electricity generation is still from fossil fuels, the increasing use of energy will itself contribute to increase carbon emissions and ultimately climate change.

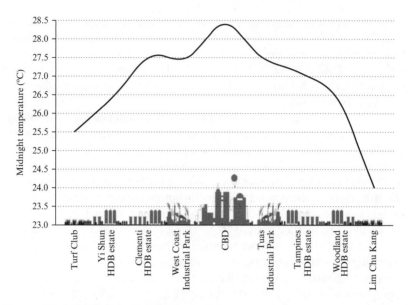

Source: Wong (2011). Reproduced with permission.

Figure 4.1   Sketch of Urban Heat Island profile in Singapore

The final direct impact is the implication of global warming on Singapore's natural landscape, which would mean a further loss in biodiversity to Singapore's already plundered natural habitat. Studies have shown how Singapore has already lost about 95 per cent of its natural habitats due to land clearing and land reclamation projects (Brook et al. 2003; Pickrell 2003; Sodhi et al. 2004) for development since independence, thus resulting in a substantial loss of Singapore's original biodiversity. Not only is there less than 5 per cent of its original mangroves left, Singapore has lost 67 per cent of its birds, 40 per cent of its animals and 5 per cent of its amphibians and reptiles, and the extinction of 39 per cent of its native coastal plants (Biodiversity Portal of Singapore 2010). There have also been instances in which the Singapore government has had to forgo some aspects of biodiversity in its efforts to ameliorate the effects of urbanization (that is, the introduction of invasive species into green spaces) or address other environmental challenges (for example, water security) – as shown in the section on strategies to address direct implications of climate change.

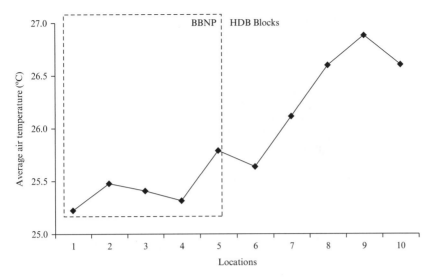

Note: Temperatures recorded in locations within BNPP (that is, points within the dotted square) and directly surrounding BNPP have lower average temperatures than other highly urbanized areas with limited green spaces.

Source: Wong (2011). Reproduced with permission.

Figure 4.2   *Temperatures of locations in close proximity to Bukit Batok Nature Park (BBNP)*

4.1.2 Indirect Impacts

Perhaps the most pressing indirect effect of climate change for Singapore is the issue of food security. This is because Singapore imports more than 90 per cent of its food from other countries. As of 2008, food imports into Singapore came from 31 countries (Hangzo 2010). The high dependence on food import is not only due to the rapid population growth and increasing food demands in Singapore, but also partly the process of urbanization since the 1970s, which has reduced the available land area devoted to farmland. The impact of climate change, however, is only one of several factors that can affect food security. Teng and Escaler (2010) highlight the various dimensions in examining food security – from macroeconomic issues relating to the availability (agricultural productivity) and physical access (market supply chain) of food to household economic access and utilization of food. Nevertheless, there have been recent incidents that demonstrate how climate-related disasters in countries providing food sources can have an impact on Singapore's food security. The severe floods in Queensland, Australia in 2011 as a result of heavy rainfall, for instance, resulted in extensive damage to various farmlands in the state and ultimately caused disruption to some of Singapore's main food imports, namely pork, beef, mutton and sugar. Such a sudden disruption in the food supply would cause a hike in food prices, therefore affecting consumers. If this incident can happen in Australia – a developed country with generally competent levels of disaster capacity – how then can poorer countries cope with the impact of climate change on their food production? As such, regional and global environmental security concerns are also a concern for Singapore's security.

In light of the direct and indirect impacts posed by climate change, which are further complicated by other factors such as urbanization and market forces, there is a need for meticulous strategies that address the various nuances and perspectives embedded in these issues. The two following sections highlight strategies that have been adopted by the Singapore government to address the direct and indirect implications of climate change, which are limited by the fact that the emphasis on top-down technical solutions ignores, to some extent, possible solutions from the bottom up.

4.2 INFRASTRUCTURAL EXCELLENCE, INSUFFICIENT RESILIENCE

It must be noted that while aspects of urbanization have exacerbated the direct impacts of climate change in Singapore, the meticulous urban planning policies during Singapore's early development phase have also provided a solid foundation and resource capacity to address the effects of climate change. Indeed, these policies would not have been possible without the support of sound economic policies in the 1960s to the 1980s that sought to address unemployment and maintain Singapore's competitive edge. The subsequent infrastructure and technological measures have thus been the bedrock of Singapore's defences against weather-related disasters, and are indeed factors that have not been fully adopted in other Asian countries. That said, however, incidents such as the Orchard Road floods have demonstrated that these infrastructural and technological measures cannot be the only crutch and must be complemented with more proactive initiatives from the community level.

4.2.1 Existing Urban Development Plans as a Foundation

Singapore's existing urban development policies have provided a solid foundation for CCA, namely in the following aspects: (1) overcoming water scarcity; (2) maintaining high levels of sanitation to avoid the spread of diseases; (3) introduction of greenery via Singapore's Garden City policies; and (4) reducing flood-prone areas in Singapore's urban environment.

Addressing water scarcity
In terms of addressing water scarcity, it should be noted that securing Singapore's water supplies was initially not driven by the effects of climate change but rather by the fact that Singapore is a highly water-stressed country (Lee 2005) and that its last water treaty with Malaysia that secures imported water will expire in 2061. Diversifying Singapore's water supply sources has thus been a primary area for much investment in research and technology. Singapore has four main water supply sources: (1) imported water (from Malaysia); (2) desalination, for which Singapore houses one of Asia's largest seawater reverse-osmosis plants; (3) water recycling (known as NEWater); and (4) water catchment areas.

There are ongoing plans to increase the capacity of the last three sources of water. While NEWater and desalination, respectively, meet about 30 per cent and 10 per cent of Singapore's current water needs, the Singapore government is working to increase the capacity of these two technologies, so that they will meet up to 50 per cent and 30 per cent, respectively, of Singapore's water needs by the year 2060 (PUB 2010a). Singapore's urban planning blueprints have included how best to maximize and enlarge Singapore's water catchment areas. The most prominent project to date is the Marina Barrage, which is built at the south of Singapore where the Singapore River meets the open sea, thus making it Singapore's largest man-made reservoir in the heart of the city.

Maintaining sanitation and mitigating diseases
Singapore has constantly ensured a high degree of sanitation thanks to several policies. Singapore's urban housing policies since the 1960s, for instance, were designed to address the lack of sanitation and mitigate the spread of communicable diseases in rural areas that existed at that time, as well as to gradually meet the needs of an increasing population. The meticulous public housing planning in Singapore also ensured easy access to amenities in various neighbourhoods, thereby avoiding overcrowding in the city centre. These policies have indeed paid off as Singapore has avoided the formation of urban slums, which are an apparent feature in many other cities in Asia (Ooi and Phua 2007) and have often been sites that are most vulnerable to climate disasters (UN-HABITAT 2008, 2011; UN-HABITAT and UNESCAP 2010; UNISDR 2010; WWF 2009).

Despite being a highly urbanized city-state, Singapore has also ensured proper waste management and sanitation measures, which are critically important for a densely populated city. In this regard, Singapore has passed strong laws and implements them through the National Environmental Agency (NEA) under the Ministry of Environment and Water Resources (MEWR). The NEA vigorously implements measures to ensure public cleanliness and hygiene, the monitoring of land and air pollution, the management of solid waste and the prevention and control of vector-borne diseases. The NEA has been particularly active in preventing and monitoring the spread of dengue fever in public areas and households, namely in increasing public awareness on clearing potential breeding spots for mosquitoes such as stagnant water in potted plants and drains. These preventive action and surveillance capacities are critical to securing high environmental health standards in Singapore.

Increasing urban greenery
The introduction of green spaces as early as the 1980s has not only been a means of replanting the loss of greenery due to urbanization but also served, in part, to alleviate the UHI phenomenon. Recent CCA policies that build on this include the Building and Construction Authority's (BCA) Green Mark Scheme and the Singapore Green Building Product Certification, which were introduced in 2005 and 2010, respectively. Both these schemes were established to encourage environmental sustainability in the design of Singapore's building and development projects, and contribute to the government's target of 'greening' 80 per cent of all buildings in Singapore by 2030. Incentives under the Green Mark Scheme for businesses to adopt environmentally-friendly practices in development projects include rewarding construction companies with work to build extra floors if they incorporate green spaces and energy-efficient features in the blueprints of future buildings. To further support the growth of energy-efficient buildings, the Singapore government has introduced environment-related tertiary courses to prepare students for jobs in environment-related industries, as well as research in areas such as solar energy. These policies are significant for two main reasons. First, they demonstrate an expansion of the rationale of being a Green City. Rather than previous garden city policies that focused on creating green spaces at ground level, green spaces are now introduced above ground level, such as green roofs and vertical planting to create green walls. Second, these measures expand the opportunities for the private sector to play a part in addressing environmental issues.

Reducing flood-prone areas
Existing urban development policies have reduced the amount of flood-prone areas in Singapore over time. While it is true that urbanization can

increase the likelihood of run-off waters, and thereby increase the risk of flooding, the Singapore government has consistently sought to improve the drainage systems (PUB Singapore 2011). This was particularly the case when Singapore experienced floods in the 1960s and 1970s (PUB 2012). In a 30-year span, flood-prone areas in Singapore have reduced from 3178 hectares in 1970 to 62 hectares in 2010 and are projected to be reduced even further (Hua and Loh 2011; PUB 2010b). Another recent feature in flood control for Singapore is the Marina Barrage, which manages the flow of water depending on the tides. There are also concerns over potential flooding and coastal land loss as a result of sea level rise coupled with Singapore's extensive land reclamation efforts. Over the course of more than 40 years (1965–2012), Singapore's land size increased from 518.5 square kilometres to 714.3 square kilometres (DOS Singapore 2012). Preliminary studies have nevertheless shown that flood mitigation solutions are within the Singapore government's ability (AsiaOne 2012; My Paper 2011; Siong 2012). Given the high value of Singapore's market land, the costs of protecting Singapore's developed coastline will be much less than the value of the potentially inundated land (Ng and Mendelsohn 2006) and therefore gives great reason for the Singapore government to implement solutions.

While these infrastructural policies play a significant role in being the first line of defence against the direct impacts of climate change, this is not enough. The following subsection highlights other initiatives that need to be considered to ensure that Singapore society as a whole has the ability to adapt to climate change.

4.2.2 What's Missing?

Recent events such as the Orchard Road floods have demonstrated that reliance on technical flood mitigation strategies – such as increasing drainage capacities and installation of flood barriers – is insufficient for adapting to climate change. There are three reasons for this. First, some of the infrastructure that has been adopted has been at the expense of other environmental concerns. Biodiversity, for instance, remains on the tail end of environmental priorities despite being further threatened by climate change, as seen from significant governmental projects such as the Marina Barrage. Despite all the benefits that the Marina Barrage has brought in terms of ensuring water security and flood protection, some experts have suggested that it will have adverse impacts on biodiversity, as the water within the bay will change from salty to fresh water and several species living in the bay may not survive the change in salinity (Biodiversity Portal of Singapore 2010).

Second, dependence on technical flood mitigation can possibly lead to unintended flooding elsewhere. This was evident during the 2011 flood in Orchard Road, which was in part due to flood mitigation responses after the 2010 flood. Areas that had been flooded in 2010 were repaved to a higher level, which was subsequently believed to have diverted flood waters to other areas, thus causing the flood in 2011 (MEWR 2012).

Third, infrastructural CCA efforts need to be complemented with community disaster preparedness efforts. While the Singapore government has, by and large, demonstrated its technical capacities to adapt to the effects of climate change, society's ability to adapt to climate change has been less certain. Norris et al. (2008) provide a comprehensive framework of community resilience based on various theories on resilience. They note that community resilience is based on a 'network of adaptive capacities', which is needed to facilitate four significant areas – economic development, social capital, information and communication, and community resilience. A list of the specific adaptive capacities in each of these areas is highlighted in Table 4.1. In the case of Singapore, it has channelled substantive efforts

Table 4.1   *Adaptive capacities contributing to economic development, social capital, information and communication and community resilience*

Area	Adaptive capacities
Economic development	• Fairness of risk and vulnerability to hazards • Level and diversity of economic resources • Equity of resource distribution
Social capital	• Received (enacted) social support • Perceived (expected) social support • Social embeddedness (informal ties) • Organizational linkages and cooperation • Citizen participation, leadership and roles (formal ties) • Sense of community • Attachment to place
Information and communication	• Narratives • Responsible media • Skills and infrastructure • Trusted sources of information
Community resilience	• Community action • Critical reflection and problem-solving skills • Flexibility and creativity • Collective efficacy, empowerment • Political partnerships

Source: Adapted from Norris et al.'s (2008) network of adaptive capacities, p. 136.

into adaptive capacities related to economic development and information and communication (as highlighted in the preceding sections). What have been less visible are Singapore's adaptive capacities relating to social capital and community competence.

This trend in Singapore can be explained with the following reasons. First, the predominantly top-down approach that has given an emphasis to infrastructural solutions has downplayed the importance of social bonds. As such, aspects of community competence (that is, community action, collective efficacy empowerment and political partnerships) are limited in an urban setting. Sapirstein (n.d.) supports this notion by noting that the process of urbanization creates a doctrine of independence, in which a nuclear family unit's ability to be self-sustaining and supportive is reduced. In addition, the urban environment has contributed to a reduced sense of community and attachment to place, which are vital components of cultivating social capital as an adaptive capacity. By increasing the level of social capital available, there will be a greater level of 'redundancy', which refers to means/systems of coping with disaster. 'The greater the redundancy, the greater the level of social resilience' (Sapirstein n.d., p. 5).

Community action and citizen leadership is also limited given public perceptions of disaster preparedness. In fact, the urban environment would also create a sense of infallibility, given Singapore's low risk and low vulnerability to weather-related disasters. The lack of resilience amongst Singaporeans is reflected in a recent survey by *National Geographic*, based on the responses of 500 Singapore citizens and permanent residents between the ages of 18 and 64. The survey noted that 79 per cent of Singaporeans do not consider themselves to be prepared for a major disaster. Half of the 500 respondents noted that readiness for a disaster is the responsibility of the government, 38 per cent also indicated that they did not know what they should be doing to prepare for a disaster, while 23 per cent did not think any disasters would happen (Tan 2012). This also gels with other surveys that have noted the lack of awareness amongst Singaporeans about sea level rise and a perception that the responsibility to address this issue lies with the government (Ng and Mendelsohn 2006, p. 295).

Second, the lack of experience in a disaster would make it difficult for a community to know how to adapt to crises. For Singaporeans, floods are perceived to be an occurrence in neighbouring countries such as Indonesia and the Philippines, and hence are difficult to picture happening in a highly urbanized and well-ordered country like Singapore. The Orchard Road floods, which had only been knee-high and drained off in a matter of hours, would hardly be considered a disaster by residents in Jakarta or the

Philippines, unlike the reaction of Singaporeans who have little experience with floods.

Countries that have often experienced disasters have made efforts to mainstream their CCA measures with disaster risk reduction (DRR) measures. This is evident in the Philippines' National Climate Change Act, which was launched in 2009 after the disastrous Typhoon Ketsana and subsequently led to the creation of the Philippines' Climate Change Commission and support for the Philippines' Disaster Risk Reduction and Management Act of 2010. This shows substantial progress and the Philippines' laws on DRR and CCA have received praise from the United Nations as being 'the best in the world' (Ulbac 2012). We have not, however, seen the same degree of CCA and DRR streamlining in Singapore. Hence, it could be argued that while the minimal number of disasters in Singapore reflects the strong technological and infrastructure readiness, it reduces the opportunity for community readiness to be put to the test. While there have been mock-runs of disaster situations, the intensity of the event would still be different from a real situation. This is not to suggest that we should hope for a disaster to happen in Singapore, but rather that some degree of exposure to a real disaster situation would give Singaporeans an element of resilience that is lacking.

4.2.3   Enhancing Resilience in Singapore

There are several things that can be done to boost the level of community resilience in Singapore. Drawing on Sapirstein's assertion of a culture of independence that impedes community resilience, there needs to be efforts to instil a culture of interdependence. This would be reflected in multi-sectoral collaborations between the private sector, government agencies, non-governmental organizations and other community organizations, with the focus on the needs and capacities of people rather than the needs and capacities of infrastructure.

At the national level, this culture of interdependence would need to employ effective inter-ministerial cooperative strategies, which is often termed a 'whole of government' approach. To date, disaster preparedness exercises have not fully focused on environmental disasters, given the concern with other disaster situations such as fires and terrorist attacks. Flooding, for instance, would pose a different set of concerns and it is therefore important to prepare for this, even though the risks of it happening are low.

Singaporeans can also work on increasing their resilience to weather-related disasters by exposing themselves to such adversities at the regional level. Neighbouring countries such as Indonesia and Thailand are highly prone to environmental disasters and it is unlikely that weather-related disasters will not happen in the future. As such, it would be useful for Singaporeans to participate in civil humanitarian efforts in these neighbouring countries as this would not only allow Singaporeans to be more aware of the circumstances during a flood but also be a means of enhancing the level of people-to-people interaction within the Association of Southeast Asian Nations' (ASEAN) regional cooperative efforts on disaster management. Such opportunities would build on literature that suggests that post-disaster volunteering can increase community resilience (Barker 2011), such as the emergence of *borantia* volunteers since the Kobe earthquake in Japan in 1995 (Childs 2008). It would be a win-win situation for all sides. Disaster-stricken communities would have alternative sources of relief assistance especially if their own national military capacity could not provide it, while the volunteers would increase their awareness and exposure to disaster situations.

Such arrangements, however, would require sufficient preparation, such as proper training for potential volunteers and on-site experienced mentors. Mercy Relief, a local humanitarian organization in Singapore, has been doing exceptional work in this field. In addition to organizing humanitarian expeditions, Mercy Relief launched a Diploma-Plus Certificate Programme in Humanitarian Affairs at Singapore Polytechnic in September 2010 as a means of grooming future humanitarian aid workers by drawing on Asian perspectives (Mercy Relief 2012). In the future, it would be useful to build and upscale such initiatives as part of a wider organizational arrangement to increase community resilience in Southeast Asia.

There is also a need for greater mapping of existing public–private partnerships. At the national level, community organizations need to be more aware of how to climate-proof their activities. In other words, community organizations need to consider further how to respond to their clientele during times of a disaster. Moreover, ad hoc arrangements that have spawned in times of disasters should also be looked at further to see how they can develop into something more sustainable. At the regional level, Singapore's efforts in disaster relief include the Singapore Red Cross working with the business community in Myanmar during Cyclone Nargis in 2008, and DHL's role in disaster relief assistance for the management of incoming relief aid into airports (DHL 2012). With greater people-to-people contact, it is hoped that Southeast Asian countries will be better able to appreciate their dependence on each other for development and security.

4.3   CONTROLLING FOOD SUPPLIES, UNCONTROLLABLE FOOD DEMANDS?

While the bulk of this chapter is focused on the direct impacts of climate change, it is worth noting Singapore's strategies to cope with the indirect effects of climate change, which can in fact be more difficult to control. Food insecurity is clearly a case in point, given the complex issues pertaining to it. This section sheds light on how the Singapore government is attempting to address the issue, but also raises important aspects that have not been fully addressed and can be pushed further with a bottom-up approach.

4.3.1   Securing Supply of Food

In addressing food insecurity as an indirect impact of climate change, Singapore has sought to tackle issues that may occur in food production. Singapore's three key strategies in addressing food insecurity are (1) diversifying food sources; (2) stockpiling essential food items; and (3) enhancing local production (Tan-Low 2011). In terms of diversifying food sources, Singapore has joined the global bandwagon of investing in other countries to secure food production. One such project is Singapore's investment in the China Jilin Modern Agricultural Food Zone, which is a joint venture between Singapore and China, worth 110 billion yuan (US$18.04 billion) (Chang and Kleyn 2010). The project is still a long way from materializing as the planned 'Super Farm' will take up to 15 years to complete. Other efforts include providing consumers with different products, such as frozen meats and food substitutes. Efforts are also being made into conducting feasibility studies on investments in other overseas food zones and imports from non-major exporters to Singapore.

As for boosting local production, there are initiatives to encourage Singaporeans to venture into farming, namely for the production of chicken, pork, fish, eggs, leafy vegetables or rice. These initiatives would be a means of increasing current 'local production of fish from 4 per cent of domestic demand ... to 15 per cent, eggs from 23 per cent to 30 per cent and leafy vegetables from 7 per cent to 10 per cent' (Chang 2010a). Singapore is also investing further resources into research and development for food security. This includes an investment of US$8.2 million by the National Research Foundation (NRF) over five years, a research partnership between Singapore's Temasek Life Sciences Laboratory, the National University of Singapore (NUS) and the Philippines-based International Rice Research Institute, to improve rice production (Marusiak 2011). In this regard, not only is the research and development geared at securing

Singapore's food production but also for the benefit of the region's food security.

4.3.2 Rethinking Demands on Food

In terms of enhancing food security, while the Singapore government has put substantial efforts into ensuring a steady supply of food, more can be done to improve or control the public's demand for food. While there has largely been an emphasis on ensuring sustained food production, reducing food wastage is just as important, currently making up about 10 per cent of the total waste generated in Singapore (NEA Singapore 2011). As such, awareness of food production cycles and a conscious effort to operationalize environmental ethics in household consumption patterns – such as reducing food waste and buying locally produced or regionally produced food – as well as firm regulations in the food and beverage industry to reduce food waste would translate into a more efficient use of food resources overall. This would perhaps be related to the dimensions of access and utilization of food. There are several ways in which people are responding to this. The first is more awareness of one's patterns of consumption. The vegetarian movement, for instance, attempts to highlight the benefits of being vegetarian and reducing the resources needed to produce meat. Another is to look to initiatives such as the UK's Waste Resources and Action Programme, which engages supermarkets to adopt measures to increase public awareness of the food products' shelf life (Chua 2012). Though these small actions may seem trivial to some critics, the potential total food waste reduction and possible reduced demand for food could be substantial if all households in Singapore are proactive in participating in them.

There is also a need to discuss the quality of the food that is being imported. For instance, with increasing concern to produce more food, genetically modified (GM) goods may be considered by the Singapore government as an option to ensure the steady supply of food for Singapore. In an interview with the *Straits Times*, Dr Ngiam Tong Tau, former Chief Executive Officer and Director General of the Primary Production Department of the AgriFood and Veterinary Authority (AVA), noted that Singapore does allow GM food, mainly corn, soya beans and products derived from them (Chang 2010b). According to a Genetic Modification Advisory Committee survey (GMAC 2007), there is a slightly greater acceptance by Singaporeans of GM foods compared to its previous survey conducted four years earlier. Nevertheless, the negative impact of GM foods and the results of the survey have been questioned by members of Singapore's environmental community (Prakash 2010). The issue has also been raised in Parliament by Ms Faizah Jamal, a member of the Nature Society and a Nominated Member of Parliament. In response, the government has noted the concerns but also highlighted the complexities in addressing GM foods, which include the possible negative impacts on food trade and food prices, as well as cross-border enforcement challenges (MND Singapore 2012).

There are also efforts to highlight the nuances of food security issues in Singapore such as the inequitable access to food within Singapore itself. In this regard, the weekly vegetarian Soup Kitchen Project has been operating since 2009 to provide food to needy folk living around the Little India precinct who have little access to decent meals. Beneficiaries of the Soup Kitchen Project range from poor senior citizens to unemployed foreign labourers, who are served by volunteers through the Food #03 network (The Soup Kitchen Project 2012). The project has been gaining momentum since 2009 and recent volunteers include batches of young students. It is an innovative yet grounded way of creating holistic awareness on the issue of food security.

4.4   CONCLUSION

Taking these various developments into consideration, it is possible to conclude that economic development – and the subsequent process of urbanization – has been a double-edged sword in Singapore's capabilities for adapting to climate change. On the one hand, economic development has provided Singapore with a strong financial base to develop a range of infrastructural and technological measures to address the effects of climate change, on top of the country's strong regulatory processes and centralized urban planning system. On the other hand, the emphasis on economic and infrastructural development has sidelined the importance of social capital and community competencies, which are essential for building resilience and disaster preparedness at the community level. Moreover, while Singapore's risk of climate-related disasters is relatively less severe than its neighbours in Southeast Asia, urbanization can amplify the effects, as seen from the Orchard Road floods.

Urban development will certainly continue in Singapore given the Singapore government's plans to prepare for a larger population – a projected 6.5 to 6.9 million by 2030 (PMO Singapore 2013) – to sustain the country's economic growth. It remains to be seen, however, whether the urban development will be able to accommodate the needs of a growing population (for example, food, water and sanitation) as well as ensuring that its technical flood mitigation mechanisms are updated with the latest scientific findings on climate change and how they relate to wider socio-economic concerns in Singapore. There is certainly scope and potential to build bottom-up community resilience adaptive capacities and forge more behavioural change amongst Singaporeans to prevent societal complacency towards disaster preparedness and dependence on infrastructural mechanisms to adapt to the effects of climate change. These national efforts must be complemented with increased regional collaborations at the local level to sensitize Singaporeans to the realities of the effects of climate change, such as the 2011 and 2013 floods in Bangkok and Jakarta, respectively.

Such developments thus reflect that in order for Singapore to effectively adapt to both direct and indirect effects of climate change, wider impacts and responses of climate change in the region must be considered and integrated into Singapore's national- and community-level strategies. Failure to do so would potentially waste viable forms of social capital that are just as important as financial capacities in improving Singapore's adaptive capacities towards climate change.

REFERENCES

AsiaOne (2012), 'Bukit Timah diversion canal expanded for better flood protection', *AsiaOne*, 20 September, available at http://www.asiaone.com/News/Latest%2BNews/Singapore/Story/A1Story20120920-372703.html (accessed 24 August 2012).

Barker, H. (2011), 'Flooding in, flooding out: how does post disaster volunteering build community resilience?', Masters Dissertation, Lund University.

Biodiversity Portal of Singapore (2010), 'Singapore, an interesting case study', in *The Diversity of Life on Earth: From Heritage to Extinction*, available at http://www.biodiversity.sg/assets/Uploads/TheDiversityOfLifeOnEarth-Chapter8.pdf (accessed 24 August 2012).

Brook, B.W., N.S. Sodhi and P.K.L. Ng (2003), 'Letters to Nature', *Nature*, **424**, 420–6.

Chang, A.L. (2010a), 'Ensuring Singaporeans don't go hungry: ex-AVA chief working on project to create vast food bowl in China', *The Straits Times*, 9 June, available at http://wildsingaporenews.blogspot.com.au/2010/06/ensuring-singaporeans-dont-go-hungry.html (accessed 24 August 2012).

Chang, A.L. (2010b), 'Of cats and dogs', *The Straits Times*, 9 June, available at http://wildsingaporenews.blogspot.com.au/2010/06/ensuring-singaporeans-dont-go-hungry.html (accessed 24 August 2012).

Chang, A. and G. Kleyn (2010), 'Singapore secures China as future food source', Strategic Analysis Paper, Future Directions International, 24 September, available at http://www.futuredirections.org.au/files/1285301165-FDI%20Strategic%20Analysis%20Paper%20-%2024%20September%202010.pdf (accessed 24 August 2012).

Childs, I. (2008), 'Emergence of new volunteerism: increasing community resilience to natural disasters in Japan', in K. Gow and D. Paton (eds), *The Phoenix of Natural Disasters: Community Resilience*, New York: Nova Science, pp. 171–80.

Chua, G. (2012), 'Cut food waste, redistribute extra food, say experts', *The Straits Times*, 21 July, available at http://www.zerowastesg.com/2012/07/31/cut-food-waste-redistribute-extra-food-say-experts-news/#more-1696 (accessed 8 August 2012).

DHL (2012), 'DHL contributes to aid efforts for Myanmar cyclone victims', Press Release, 18 May, available at http://www.dp-dhl.com/en/media_relations/press_releases/2008/aid_efforts_for_myanmar_cyclone_victims.html (accessed 8 August 2012).

DOS (Department of Statistics) Singapore (2012), *Statistics Singapore – Key Annual Indicators*, available at http://www.singstat.gov.sg/stats/keyind.html#popnarea (accessed 24 August 2012).

GMAC (Genetic Modification Advisory Committee) (2007), 'Survey indicates Singaporeans' knowledge and attitudes towards genetic modification have improved slightly since 2001', News Articles, 25 January, available at http://www.gmac.gov.sg/News/2007/2007_01_25_GMAC.html (accessed 24 August 2012).

Grosse, S. (2012), 'Singapore scientists find dengue killer in human antibody', Channel News Asia, 21 June, available at http://www.channelnewsasia.com/stories/singaporelocalnews/view/1208989/1/.html (accessed 8 August 2012).

Hangzo, P.K. (2010), 'Facing food shortages: urban food security in an age of constraints', RSIS Commentaries, 17 August, available at http://www.rsis.edu.sg/publications/Perspective/RSIS0922010.pdf (accessed 24 August 2012).

Hua, D. and Y.W. Loh (2011), 'Managing storm water in urbanised Singapore for flood control', Paper presented at Flood and Storm Surge Control Training Meeting, Tokyo, 25–27 January, available at http://www.asianhumannet.org/db/datas/1102/singapore.pdf (accessed 19 June 2012).

Khasnis, A.A. and M.D. Nettleman (2005), 'Global warming and infectious disease', *Archives of Medical Research*, **36**, 689–96, available at http://www.bvsde.paho.org/bvsacd/cd68/AKhasnis.pdf (accessed 19 June 2012).

Lee, P.O. (2005), 'Water management issues in Singapore', Paper presented at the Water in Mainland Southeast Asia Meeting, 29 November–2 December, Siem Reap, Cambodia, available at http://www.khmerstudies.org/download-files/events/Water/Lee%20Nov%202005.pdf (accessed 19 June 2012).

Marusiak, J. (2011), 'Singapore to address global food security through R&D', *EcoBusiness*, 15 August, available at http://www.eco-business.com/features/singapore-to-address-global-food-security-through-rd/ (accessed 19 June 2012).

Mercy Relief (2012), Academic Training Courses, available at http://www.mercy relief.org/web/Contents/Contents.aspx?ContId=304 (accessed 19 June 2012).

MEWR (Ministry of Environment and Water Resources) Singapore (2009), *Singapore's National Climate Change Strategy*, available at http://app.mewr.gov.sg/data/ImgUpd/NCCS_Chapter_2_-_VA.pdf?&lang=en_us&output=json (accessed 19 June 2012).

MEWR (Ministry for the Environment and Water Resources) Singapore (2012), *Expert Panel Report on Drainage Design and Flood Protection Measures*, available at http://app.mewr.gov.sg/data/ImgCont/1524/Expert_Panel_Report_on_Drainage_Design_and_Flood_Protection_Measures.pdf (accessed 19 June 2012).

MND (Ministry of National Development) Singapore (2012), *Written Answer by Ministry of National Development on the Reviewing of Genetically Modified Food*, 14 May, available at http://app.mnd.gov.sg/Newsroom/NewsPage.aspx?ID=3554&ca

tegory=Parliamentary%20Q%20&%20A&year=2012&RA1=&RA2=&RA3= (accessed 19 June 2012).

My Paper (2011), 'Rules changed to improve flood-prevention measures', *My Paper* (Singapore), 1 December, available at http://news.asiaone.com/News/ AsiaOne%2BNews/Singapore/Story/A1Story20111201–313693.html (accessed 24 August 2012).

NEA (National Environment Agency) Singapore (2011), *Waste Statistics and Overall Recycling*, available at http://app2.nea.gov.sg/topics_wastestats.aspx (accessed 19 June 2012).

Ng, W.S. and R. Mendelsohn (2005), 'The impact of sea level rise on Singapore', *Environment and Development Economics*, **10**, 201–15.

Ng, W.S. and R. Mendelsohn (2006), 'The economic impact of sea-level rise on nonmarket lands in Singapore', *Ambio*, **35** (6), 295.

Norris, F.H., S.P. Stevens, B. Pfefferbaum, K.F. Wyche and R.L. Pfefferbaum (2008), 'Community resilience as a metaphor, theory, set of capacities, and strategy for disaster readiness', *American Journal of Community Psychology*, **41**, 127–50.

Ooi, G.L. and K.H. Phua (2007), 'Urbanization and slum formation', *Journal of Urban Health*, **84** (1), May, 27–34.

Patz, J.A. and S.H. Olson (2006), 'Malaria risk and temperature: influences from global climate change and local land use practices, in *Proceedings of the National Academy of Sciences of the United States of America (PNAS)*, **103** (15), 5635–6, available at http://www.pnas.org/content/103/15/5635.long (accessed 19 June 2012).

Pickrell, J. (2003), 'Singapore extinctions spell doom For Asia?', *National Geographic News*, 23 July, available at http://news.nationalgeographic.com.au/ news/2003/07/0723_030723_singapore.html (accessed 24 August 2012).

PMO (Prime Minister's Office) Singapore (2013), *A Sustainable Population for a Dynamic Singapore: Population White Paper*, available at http://population.sg/ (accessed 31 January 2013).

Popatlal, A. (2009), 'Thursday's floods an event that occurs once every 50 years', Channel News Asia, 20 November, available at http://www.channelnews asia.com/stories/singaporelocalnews/view/1019504/1/.html (accessed 19 June 2012).

Prakash, B. (2010), 'Better labelling of GM foods in Singapore essential', 16 June, available at http://www.ecowalkthetalk.com/blog/2010/06/16/better-labelling-of-gm-foods-in-singapore-essential/ (accessed 19 June 2012).

PUB (Public Utilities Board) Singapore (2010a), *Water Supply – Then and Now*, December, available at http://www.pub.gov.sg/about/historyfuture/Pages/ WaterSupply.aspx (accessed 19 June 2012).

PUB (Public Utilities Board) Singapore (2010b), *Flood Management in Singapore*, available at http://www.pub.gov.sg/managingflashfloods/FMS/Documents/Flood ManagementSg.pdf (accessed 19 June 2012).

PUB (Public Utilities Board) Singapore (2011), *Overview of Singapore's Drainage Management Approach*, July, available at http://www.pub.gov.sg/general/ Documents/overview_DrainageMgmt.pdf (accessed 19 June 2012).

PUB (Public Utilities Board) Singapore (2012), *Managing Flash Floods*, available at http://www.pub.gov.sg/managingflashfloods/Pages/default.aspx (accessed 19 June 2012).

Sapirstein, G. (n.d.), 'Social resilience: the forgotten element in disaster reduction', available at http://www.oriconsulting.com/social_resilience.pdf (accessed 19 June 2012).

Siong, O. (2012), 'Detention tank, diversion canal to enhance flood prevention', Channel News Asia, July 19, available at http://www.channelnewsasia.com/stories/singaporelocalnews/view/1214405/1/.html (accessed 24 August 2012).

Sodhi, N.S., L.P. Koh, B.W. Brook and P.K.L. Ng (2004), 'Southeast Asian biodiversity: an impending disaster', *Trends in Ecology and Evolution*, **19** (12), 654–60.

Tan, J. (2012), '79% of S'poreans not prepared for major disaster: survey', 17 September, available at http://sg.news.yahoo.com/79--of-s%E2%80%99poreans-not-prepared-for-major-disaster--survey.html (accessed 24 August 2012).

Tan-Low, L.K. (2011), 'Food security and Singapore's position', Paper presented at the ICRM Symposium 2011, 3 March, available at http://icrm.ntu.edu.sg/ICRM_doc/ICRM/Documents/ICRM%20Symposium%202011/Tan-Low%20Lai%20Kim.pdf (accessed 24 August 2012).

Teng, P. and M. Escaler (2010), 'The case for urban food security: a Singapore perspective', NTS Perspectives, No. 4, RSIS Centre for NTS Studies, Singapore.

The Soup Kitchen Project (2012), available at https://www.facebook.com/pages/The-Soup-Kitchen-Project-Singapore/300766934440 (accessed 10 June 2012).

Today Online (2012), 'Urbanisation has led to increase in storm water run-off: expert panel', 11 January, available at http://www.eco-business.com/news/urbanisation-has-led-to-increase-in-storm-water-run-off-expert-panel/ (accessed 12 May 2012).

Ulbac, M.L. (2012), 'UN lauds Philippines' climate change laws "world's best"', *Phillippine Inquirer*, 4 May, available at http://globalnation.inquirer.net/35695/un-lauds-philippines%E2%80%99-climate-change-laws-%E2%80%98world%E2%80%99s-best%E2%80%99 (accessed 24 August 2012).

UN-HABITAT (2008), *State of the World's Cities 2008/2009 – Harmonious Cities*, London, available at http://www.unhabitat.org/content.asp?cid=5964&catid=7&typeid=46&subMenuId=0 (accessed 24 August 2012).

UN-HABITAT (2011), *Cities and Climate Change: Global Report on Human Settlements 2011*, London, available at http://www.unhabitat.org/pmss/listItemDetails.aspx?publicationID=3086 (accessed 24 August 2012).

UN-HABITAT and UNESCAP (2010), *The State of Asian Cities 2010/2011, Japan*, available at http://www.unhabitat.org/pmss/listItemDetails.aspx?publicationID=3078 (accessed 24 August 2012).

UNISDR (2010), *Local Governments and Disaster Risk Reduction: Good Practices and Lessons Learned*, Geneva, available at http://www.preventionweb.net/files/13627_LocalGovernmentsandDisasterRiskRedu.pdf (accessed 24 August 2012).

Wong, N.H. (2011), 'Research highlights: a study of urban heat island in Singapore', School of Design and Environment, National University of Singapore, available at http://sde.nus.edu.sg/rsh/research/research_highlightsB01.html (accessed 24 August 2012).

WWF (2009), *Mega-stress for Mega-cities: A Climate Vulnerability Ranking of Major Coastal Cities in Asia*, Gland, available at http://reliefweb.int/sites/reliefweb.int/files/resources/20AAFE6BF5CFD2DD4925766C001E27CD-wwf_nov2009.pdf (accessed 24 August 2012).

5. Assessing climate change impacts and adaptation strategies in China

Joanna I. Lewis*

5.1 INTRODUCTION

Throughout the world, communities are struggling to build and maintain resilience to extreme weather events, drought, sea level rise and other impacts of climate change. China is one of many countries around the world beginning to develop national-level policy targeting adaptation to climate change. Because provinces vary in terms of vulnerability, degree of impact and other factors, solutions are appearing at the sub-national level as well.

The way in which climate change affects China is influenced by the country's size, geography and resource endowments, as well as by such factors as its reliance on trade and political relationships with its neighbors (Lewis 2009, 2011b). In China, rapid economic growth has come at a cost to both the local and global environment (Lewis and Gallagher 2010). As the impacts of rising global greenhouse emissions are more comprehensively understood, it is becoming increasingly evident that climate change will exacerbate many of the country's existing environmental problems, such as deteriorating water quality, water scarcity, air pollution, land degradation and desertification.

This chapter reviews the observed and predicted impacts China is likely to face due to climate change, focusing on observed physical and predicted impacts on water resources and agricultural systems, impacts facing coastal economic centers and impacts that threaten public health and national security. It examines the response of the Chinese government to date, focusing on national-level plans for climate change adaptation and for improved resilience. It concludes by highlighting some of the country's most vulnerable regions, as well as some of the most promising adaptation programs underway, in order to help shape the agenda for improving such initiatives across China at both the central and sub-national levels.

5.2 ASSESSING CHINA'S CLIMATE VULNERABILITY

Human activity is altering the climate, driven primarily by a century and a half of fossil fuel combustion (Solomon et al. 2007). Carbon dioxide (CO_2) concentrations in the atmosphere reached 395 parts per million in 2013, 41 percent higher than pre-industrial levels (Keeling et al. 2013). Average global temperatures have risen by 0.85°C since 1880, and are projected to increase by another 1.5°C–4.5°C upon a doubling of CO_2 concentration (IPCC, 2013). The results of these changes are already manifest in altered weather patterns, extreme weather events, floods, droughts, glacial and Arctic ice melt, rising sea levels and reduced biodiversity (Solomon et al. 2007).

In China, temperature data based on direct measurements demonstrates an increase in nationwide mean surface temperature of 1.38°C over the past 50 years (Lin et al. 2007). According to China's 2006 *National Assessment Report on Climate Change*, nationwide average temperatures are projected to increase between 1.3°C and 2.1°C by 2020, 1.5°C and 2.8°C by 2030 and 2.3°C and 3.3°C by 2050 from the year 2000 (Government of China 2006). Now the world's largest emitter of CO_2 on an annual basis, China was responsible for 28 percent of global emissions in 2011 (Le Quéré et al. 2012). At the core of the climate change challenge is China's energy sector, the single largest source of emissions in the world. As China's energy demand has boomed, its CO_2 emissions have soared. Looking ahead, most projections put China's CO_2 emissions in 2030 in the range of 500 percent above 1990 levels (Energy Information Administration 2011). Globally, this translates to almost 50 percent of all new energy-related CO_2 emissions between 2005 and 2030, as illustrated in Figure 5.1. If China's emissions continue to grow at the rate of 8 percent per year (the average annual growth rate between 2005 and 2010), by the year 2030, it could be emitting as much CO_2 as the entire world is today.

5.2.1 Observed Physical Impacts

A synthesis report compiled by China's leading climate change scientists found that climate change is likely to 'cause significant adverse impacts on the ecosystems, agriculture, water resources, and coastal zones in China' (Lin et al. 2007, p.6). Examples of regional impacts predicted include freezing rain and snow in the south, droughts across the middle and lower reaches of the Yangtze River and serious flooding in Beijing.

Many of these predicted impacts are already being observed. For example, drought in Yunnan province in China's southwest in early 2013

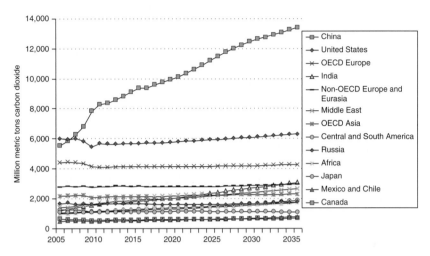

Note: Includes CO$_2$ emissions from fossil fuels only. Reported data extend through 2010; 2011–2035 emissions are from EIA's reference case projections.

Source: EIA (2011).

Figure 5.1 Carbon dioxide emissions projections by country or region

left 1.43 million people and 796,000 livestock animals without drinking water, and 55,333 hectares of farmland yielded no crops in this year (Xinhua 2013). Due to lack of precipitation water is reportedly no longer flowing in 134 medium and small rivers in Yunnan, and 138 small reservoirs have dried up, leading provincial officials to allocate 230 million yuan (US$37 million) to drought relief funds including digging wells and pumping water (Xinhua 2013). In the summer of 2012 in Beijing, extreme flooding events affected 800,000 people, killing 77 people (Whiteman 2012). During this period, about 5000 hectares of farmland were flooded, killing 170,000 livestock animals and costing an estimated 6.1 billion yuan (US$955 million), while 66,000 homes, 50 bridges and 750 kilometers of road were also damaged (Whiteman 2012).

According to a recent report by China's State Oceanic Administration, China's sea level hit a record high in 2012, aggravating tides, monsoon flooding, shoreline erosion and crop inundation, and putting homes and livelihoods at risk (Wang 2013). In 2012 alone, China experienced 24 storm surges, causing economic losses of around 13 billion yuan (US$2 billion), which is about 32 percent higher than the average damage over the preceding five years (Wang 2013).

Regional differences in precipitation have also changed over the past century, with rainfall gradually declining in the water-scarce north at a rate of 20–40 millimeters per decade and rising in the south at a rate of 20–60 millimeters per decade (World Bank 2012b). Decreased precipitation can also stem from increased desertification, which is a problem particularly in northern China. One study estimates that approximately 150 square kilometers of cultivated land is lost each year due to desertification (Kim et al. 2006).

5.2.2   Predicted Physical Impacts

While climate impacts are already being observed across China, it is the predicted future impacts that are even more of a cause for serious concern among China's inhabitants and policymakers alike.

Water resources
China's water resources are particularly vulnerable to climate change, and predicted changes are expected to exacerbate water scarcity. In northern China, where water scarcity is already a very severe problem, river runoff is expected to decrease; while in southern China, where flooding and heavy rains are already a problem, runoff is expected to increase. Modeling results show that in the coming 50–100 years, multi-year average runoff will decrease by 2 percent to 10 percent in the north and increase by 24 percent in the south (Liu et al. 2012). Flooding events previously expected to occur once in 50 years are expected to occur once in five to 20 years by 2050 (Xinhua News Agency 2011). Future flood events over most parts of China are projected to occur more frequently, and with a stronger intensity and a longer duration than currently (Chen et al. 2012). Since the 1950s, the six largest rivers in China have received declining rates of runoff, resulting in some rivers even facing intermittent water flow (Lin et al. 2007). The annual amount of irrigation water required in the Huang-Huai-Hai river plain is expected to increase by 66 to 84 percent if there is a 20 percent decrease in annual rainfall (Wang and Zhang 2011).

The impact of climate change on the glaciers in China's Tibetan Plateau will have stark ramifications for China's watersheds. The aggregate area of China's western glaciers is projected to decline by 27.2 percent by 2050 (ANI 2010). Thawing may initially increase water discharge to mountain and highland lakes, but water discharge will then decline as the glaciers recede. This phenomenon could also result in a short-term increase in water flowing to the Yellow and Yangtze River deltas, followed by a decline in river flow and eventual drought across some of China's richest agricultural regions (Economy 2007).

Approximately 17 percent of China's electricity production is from hydropower, including both small and large dams (NBS 2011). As climate change affects China's water resources, this will affect the availability of water for hydroelectric dams, particularly if river flow declines in southwestern China. If hydropower resources are reduced, this could lead to an increased dependence on fossil fuels, resulting in further increased greenhouse gas emissions. This could also potentially lead to increased Chinese reliance on imported energy resources, creating continued pressure on world oil, coal and gas markets (Lewis 2009).

Agriculture
China's agricultural system is particularly vulnerable to future climate impacts. China must feed about one-fifth of the world's population with just 7 percent of the world's arable land (Wu 2008). The world's largest producer of rice and wheat, China produced 187.5 billion tons of rice and 104.6 billion tons of wheat in 2010 (FAO 2012), primarily for domestic consumption.[1] China is also among the principal sources of tobacco, soybeans, peanuts and cotton (Poole and Ruitenberg 2011; US Department of Agriculture 2009). Climate change is expected to cause larger variations in crop productivity, and make crops more susceptible to pests and disease, as well as droughts and floods, thereby decreasing the overall stability of agricultural production (Piao et al. 2010).

China's *National Assessment Report on Climate Change* estimated a decline of up to 37 percent in China's rice, maize and wheat yields after 2050 as a result of climate change, and a 5–10 percent decline in overall crop productivity in China by 2030 (Government of China 2006; Lin et al. 2007). This scale of productivity decline poses a severe challenge for long-term food security in China, and China's connections to global agricultural trade could also have global repercussions. If this decline in supply resulted in global scarcity and elevated food prices, it could have particularly severe impacts in regions already affected by food security, especially in parts of Africa (World Economic Forum 2008). Estimates of the climate-related impacts on crop yields vary dramatically, however, depending on assumptions made about the fertilization effect of CO_2. For example, one study that examined the yield impacts on China's three major food and feed crops (rice, wheat and maize) ranged from –22.8 percent for irrigated maize to +25.1 percent for irrigated wheat by 2050 (Wang et al. 2010).

Higher CO_2 levels in the atmosphere may increase crop growth and yield through its effect on photosynthetic processes, while higher temperatures generally have the opposite affect: decreasing yield by speeding up a plant's development so that it matures quicker, as well as increasing a plant's water demand. Higher temperatures may also increase the pervasiveness of pests, diseases and weeds.

China's agricultural system is also vulnerable to encroaching desertification. China's existing desertification has been driven by climatic change and strong wind regimes, and accelerated by development and urbanization. Estimates of China's total land terrain touched by desertification range from one-third to one-fifth of total land area. At present, more than 60 percent of China's arid and semi-arid regions are managed using traditional pastoral and agricultural systems and support the livelihoods of about 80 million people (Wang et al. 2008; Wang et al. 2009). Annual direct economic loss caused by desertification is approximately US$6.5 billion (People's Daily Online 2008; Secretariat of the China National Committee 2000). In addition, as more and more of China's land is lost to urbanization, agricultural land is increasingly scarce, making any losses in yield even more problematic.

China's agricultural production was responsible for 10 percent of its total gross domestic product (GDP) in 2009 (NBS 2011), and therefore declining yields could have implications for overall economic growth as well as revenue from trade in agricultural products (Peng et al. 2004). The total value of exported agricultural commodities in 2010 was US$41.1 billion, up from US$17 billion in 2004 (NBS 2011).

Sea level rise
Since the 1950s, sea level rise has been observed along China's coastline at a rate of 0.0014 to 0.0032 meters per year (Lin et al. 2007), and recent rates observed in the South China Sea (0.0055 meters per year) are significantly higher than the rate of global sea level rise (0.0033 meters per year) (Feng et al. 2012). Estimates of future sea level rise along China's coastline range between 0.01 and 0.16 meters by 2030, and between 0.4 and 1 meter by 2050 (Fan and Li 2006; Li et al. 2001). Higher sea levels amplify the likelihood of flooding and intensified storm surges, and also aggravate saltwater intrusion and coastal erosion. Most of Shanghai is 3–4 meters above sea level (about 10–13 feet), and in close proximity to water. Other cities that are already being affected by sea level rise includes Suizhong, in Liaoning, which has experienced a retreat in its shoreline of about 60 meters since 2000, and in nearby Panjin, agricultural land about 18 km inland from the coastline was inundated by seawater resulting in soil salinization (Wang 2013).

China's 12 coastal provinces contain about 43 percent of its population, contribute about 65 percent of its GDP and have a collective per capita GDP of about 50 percent higher than the national average (Chovanec 2011; NBS 2011). They also contain several special economic zones, which receive special incentives to encourage economic activity and to target foreign investment, as well as several designated coastal development

Table 5.1   *Economic indicators and projected sea level rise in China's coastal river deltas*

Region	Nearby economic centers	Estimated sea level rise by 2050 (meters)	Population (percent of national total)	GDP (percent of national total)
Yellow River Delta (Huanghe)	Beijing, Tianjin, Shandong Province	0.7–0.9	9.6	15.6
Yangtze River Delta (Changjiang)	Shanghai, Jiangsu Province, Zhejiang Province	0.5–0.7	11.6	21.5
Pearl River Delta (Zhujiang)	Guangdong Province (Guangzhou, Shenzhen, Zhuhai), Hong Kong	0.4–0.6	8.3	14.9

Source: Sea level rise data reported in Fan and Li (2006); population and GDP data estimated based on NBS (2011).

zones and high-tech zones. As a result, China's core economic development infrastructure is directly vulnerable to the impacts of climate change, posing a great risk to China's economic stability that is fundamental to the Communist Party's own longevity.

China contains roughly 144,000 square kilometers of low-lying coastal lands – a land area equivalent to the size of the country of Bangladesh. With an elevation no greater than 5 meters, these low-lying regions are mainly distributed across the deltas of the Yellow, Yangtze and Pearl Rivers, and include the densely populated, major industrial centers of Tianjin, Shanghai and Shenzhen/Guangzhou (Fan and Li 2006). The population of China's urban coastal regions is growing at three times the national average rate, with the trend of urban migration expected to continue for decades to come (McGranahan et al. 2007). Estimates of sea level rise projected in China's coastal river deltas are presented in Table 5.1, along with the nearby economic centers, and estimates of total population and percent of GDP potentially vulnerable to this sea level rise.

The South China region, which includes the low-lying Pearl River Delta region, is susceptible to sea level rise. Sea level rise threatens the continued rapid economic growth of these low elevation coastal regions, which not only surround the economic centers of Shanghai, Tianjin and Guangzhou but are also home to key ports through which imports and exports pass, connecting China to the world. About 14 percent of China's freight goes through Shanghai, and 8 percent through Tianjin, while 29 percent of China's foreign trade income comes from Guangzhou (NBS 2011; Xinhua News Agency 2008).

Extreme weather events
Climate change is expected to increase the frequency and severity of extreme weather events such as storm surges, flooding and typhoons. Both the frequency and the intensity of tropical storm surges have increased in China since the 1960s (Fan and Li 2006). Severe flooding events are also predicted to increase in frequency, particularly in eastern China.

Shanghai is especially vulnerable to the impacts of climate change. Surrounded on three sides by bodies of water (the Yangtze River Estuary to the north, Hangzhou Bay to the south and the East China Sea to the east), Shanghai has frequently been plagued by typhoon storm surges, and its low elevation makes it particularly susceptible to sea level rise (Wang et al. 2012). A recent study of coastal vulnerability found that Shanghai was the most vulnerable city in the world to coastal floods induced by climate change as determined by a variety of hydrogeological, social and economic metrics (Balica et al. 2012). Another study estimates that by 2100, 'half of Shanghai is projected to be flooded, and 46 percent of the seawalls and levees are projected to be overtopped' (Wang et al. 2012, p. 537).

China ranked first in 2010 among all countries with the highest percent of urban populations in vulnerable coastal zones, followed by India, Japan, Indonesia and the United States, with Guangzhou and Shanghai in the top five of all cities (Nicholls et al. 2007; World Bank 2010b). Extreme weather events have already taken a toll on China's eastern coastal region; in 2006, such disasters cost China more than US$25 billion in infrastructure damage, and future exposed assets have been estimated to potentially amount to as much as US$35 trillion by 2070 (Nicholls et al. 2007).

China's major trading partners that rely on exports originating in the major eastern coastal economic centers may be affected by any major climate-related impacts, such as severe weather events. China's economic welfare is intrinsically tied to the other major economies, and therefore the economic impact of climate change on China has potential implications for countries connected to China's economy. Shanghai, particularly vulnerable to impacts associated with increased sea level rise, is China's leading commercial center with one of the world's largest ports. Shanghai's major trading partners include the United States, the European Union and Japan, and it was the source of about US$437 billion in foreign trade in 2012 (Wu and Yu 2013).

Public health

Climate change will likely influence the distribution, incidence and transmission of vector-borne infectious diseases. Climate cycles have been shown to cause inter-annual variations in disease incidence (McMichael et al. 2004), and the increased spread of vectors of tropical diseases such as malaria, dengue and Chagas disease into previously temperate regions of the world could result in a greater incidence of ill-health (Arnold et al. 2007). Malaria, one of the most harmful vector-borne diseases in the world and prevalent in parts of southern China, is expected to have an expanded geographical range into the temperate and arid parts of the country (World Health Organization 2003). In addition, viruses affecting migratory bird populations, such as strains of avian flu, may spread as climate change shifts habitats and migratory patterns of birds (Economy 2007).

After a severe flooding event, the incidents of infectious diarrhea diseases, such as cholera, dysentery, typhoid and paratyphoid, are often increased, as stagnant water encourages the spread of disease (WICCI: Human Health Working Group 2009). Another direct climate-related health impact concerns the increase in hot weather. Heat stroke particularly affects the elderly and very young, with heat-related illness and mortality reported in urban cities that experience severe heat waves. Temperature also influences the incidence of cardiovascular and respiratory diseases.

Ethnic tensions

As climate-related stresses influence human livelihoods, populations in affected regions may opt to migrate to less affected parts of the country. Some of the regions that are most vulnerable to the impacts of climate change in China are already subject to ethnic tensions, including the Tibetan Plateau region. As glaciers melt and inhabitants of the Tibetan Plateau region must deal with more water scarcity, they may migrate toward central or eastern China. A similar phenomenon may occur as Inner Mongolia faces more desertification as a result of climate change. Migration of such 'climate refugees' from Tibet and Inner Mongolia into predominantly ethnic Chinese regions could increase ethnic tensions, and even threaten regional stability.

Alternatively, populations from regions where water is scarce may move toward northwest and southwest China where rainfall and glacier runoff may result in increased water availability. The movement of Han Chinese into resource-rich Muslim Uighur and ethnic Tibetan areas may aggravate existing ethnic tensions in these regions, many of which are caused by ongoing resource extraction from the west for the east (Podesta et al. 2007).

While it is difficult to understand the exact nature of China's climate change-related impacts, it is clear that the nation's agricultural systems and

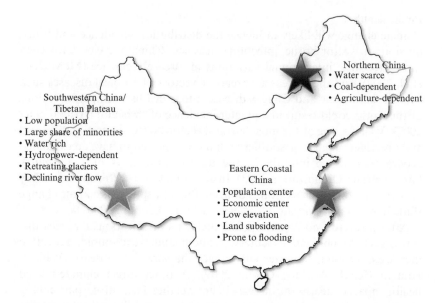

Source: Illustration by author. Representation only, territorial boundaries may not be accurate.

Figure 5.2   China's regional climate vulnerabilities

economic development engines are all vulnerable, with direct implications for human livelihoods. As these systems are all interconnected, feedbacks are highly likely. For example, if changing weather and precipitation patterns affect agricultural productivity, this in turn affects economic output and trade, as well as farmers' livelihoods. If rising sea level harms economic development along China's eastern coastline, this will directly affect global trade flows. If populations migrate either inland to flee rising sea levels and storms or from the west to the east as water resource availability changes, this may in turn influence agricultural patterns, or spur inter-regional or ethnic conflict. Additionally, certain regions across China are exposed to multiple vulnerabilities, as illustrated in Figure 5.2. As a result, China will most certainly face new challenges in a climate-changed world.

5.3   ASSESSING CHINA'S ADAPTATION RESPONSE

There is a range of options that China can begin to implement now in order to avoid some of the most significant impacts of climate change. In urban areas, there are a variety of measures that can be taken to build more resilient cities. In rural areas, particularly those that rely on agriculture, actions to improve water conservation and drought resistance of crops may have near-term benefits.

While many of China's urban and rural areas are moving forward with plans to increase resilience, real questions remain among government leaders about the true extent of vulnerability. A better characterization of the risks they face would allow policymakers to better understand where climate adaptation strategies should rank in terms of other policy priorities. Since climate change is expected to exacerbate many of China's existing vulnerabilities, including water scarcity, increasing desertification and flooding of low-lying coastal cities, taking action to address these problems has an immediate benefit no matter how severe the added impact of climate change.

China's leaders increasingly acknowledge reports from the scientific community about the impacts of climate change, and plans for climate change adaptation measures are increasingly making their way into high-level strategic policy documents. Back in 2007, China's former Special Ambassador for Climate Change stated that 'climate change . . . is in fact a comprehensive question with scientific, environmental and development implications and involves the security of agriculture and food, water resource, energy, ecology and public health and economic competitiveness' and 'if the climate changes dramatically, the survival of mankind and the future of earth might be impacted' (Yu 2007). China's first *National Climate Change Program*, released in June 2007, highlighted the government's plans for adapting to climate change (National Development and Reform Commission 2007). While the document was primarily focused on climate change mitigation, the section on adaptation highlighted plans for (1) the agricultural sector; (2) forests and other natural ecosystems; (3) water resources; and (4) coastal zones and coastal regions. In November 2012, the government released a 'white paper' called 'China's policies and actions for addressing climate change' that expanded substantially on the 2007 Plan, including elaborations on adaptation plans for public health and disaster prevention, among other areas (Information Office of the State Council, PRC 2012).[2]

These plans for adaptation to climate change, as articulated by these high-level Chinese government policy documents as well as examples of sub-national actions being taken to build resilience to climate change in vulnerable parts of the country, are examined in the following subsections. When data is available, progress to date in implementing these plans for climate adaptation, as well as the role of international support in implementing projects related to building climate resilience, is also reviewed below.

5.3.1 Promoting Agricultural Resilience and Food Security

In order to maintain domestic food security, responding to climate-related impacts affecting China's agricultural system is identified as a high priority. The government's plan for the agricultural sector includes a focus on infrastructure improvements, primarily water conservation, as well as the cultivation of seeds for agricultural crops with 'high yield potential and resistance to drought, flooding, high temperature, diseases and pests' (Information Office of the State Council, PRC 2012).

Large-scale water irrigation and drainage projects and water-saving irrigation demonstration projects are being implemented in or planned for arid and drought-prone areas. Field restoration in places that are subject to salinization and alkalization is proposed as a means of increasing land availability. Other programs include the optimization of the agricultural system structure to increase overall productivity, for example, by diversifying crops (National Development and Reform Commission 2007). The government has reported that over the 11th *Five-Year Plan* period of 2006–10, several agricultural demonstration projects designed to save water were piloted around the country, resulting in an increase in water productivity of 10–30 percent, as well as increased grain output in drought-susceptible regions (Information Office of the State Council, PRC 2012).

Plans to enhance food security also include major biotechnology developments, including the breeding of stress-resistance varieties of crops with resistance to drought, flooding, high temperature, diseases and pests (National Development and Reform Commission 2007). The government reports that more than 95 percent of China's farmland is sown with 'superior' strains, which are attributed with contributing to a 40 percent increase in grain output (Information Office of the State Council, PRC 2012).

Rice is the staple food in China, and rice production is estimated to consume between 50 and 70 percent of China's fresh water resources annually (National Development and Reform Commission 2007). As a result, drought is a key constraint in rice production in China and can result in large yield losses or limit average yields across the country (Luo 2010). Research programs supported by the Chinese government have targeted the development of 'super rices' to increase production, but even such hybrid rice varieties are highly susceptible to drought. The development of drought-resistant rice is a key topic for ongoing biotechnology research targeting dehydration avoidance, dehydration tolerance and drought recovery (Luo 2010). 'Water-saving and drought-resistant rice' (WDR) is a new rice variety developed specifically for its high yield and water-saving characteristics; conventional varieties include Zhonghan 3, Huhan 3, Huhan

15, Handao 297, Zhonghan 209; and hybrid combinations include Hanyou 2 and Hanyou 3 (Luo 2010).

Since water is already such a limiting factor in Chinese agriculture, there have been many initiatives to move toward irrigated agriculture, which may be better able to handle climate variations than rain-fed crops (Luo 2010). The World Bank has funded projects in China such as the US$55 million 'Mainstreaming Climate Change Adaptation in Irrigated Agriculture Project,' which aimed to 'enhance adaptation to climate change in agriculture and irrigation water management practices through awareness-raising, institutional and capacity strengthening, and demonstration activities' in the Huang-Huai-Hai (3H) Basin (World Bank 2012a). The 3H Basin, located on the northern China plain, has one-third of the Chinese average per capita water availability (and China's average already places it in the bottom 25 percent of countries) (World Bank 2012a). Water flows have declined significantly in the 3H Basin over the past two decades with the Huang and Huai River Basins experiencing 15 percent declines and the Hai River Basin experiencing a 41 percent decline (World Bank 2012b). As a result, water shortages are expected to increase from about 30–40 cubic kilometers per year to 56.6 cubic kilometers by 2050 (World Bank 2012b). This World Bank program specifically targeted China's largest national irrigation program, the Comprehensive Agricultural Development (CAD) Program, with a goal to 'mainstream climate change adaptation measures, techniques, and activities' (World Bank 2012b). This included the development of farmer participatory organizations to improve irrigation and water resources management, and strategies to disseminate information about adaptation approaches (World Bank 2012a).

5.3.2 Protecting Forests and Other Natural Ecosystems

While forest preservation is more directly relevant for climate change mitigation, protecting forest areas can also play an important role in increasing the resilience of natural areas. While China has a long history of destructive deforestation practices, the Forest Law of the People's Republic of China aims to protect forest ecosystems and establishes logging bans in order to allow forests to regenerate. In addition, China's State Forestry Administration recently released the *Action Points for China's Forestry Departments in Response to Climate Change During the 12th Five-Year Plan (2011–2015) Period*, which calls for intensified forest management efforts, forest fire and pest prevention, forest structure optimization and the improvement of forest health (Information Office of the State Council, PRC 2012). China's *Climate Change Program* calls for expanded conservation measures, as well as strengthening systems for fire control and disease control in forests (National Development and Reform Commission 2007). As both forest fires and diseases targeting trees are projected to increase due to climate change, these are programs that are worth implementing, though their effectiveness may be difficult to measure.

The *Regulations on Wetland Protection of the People's Republic of China* aims to stop an overall decline in wetlands. A 2011 government survey of wetland resources found that 39 wetland protection programs had been implemented nationwide, and that over 100 stations for wetland protection and management had been constructed (Information Office of the State Council, PRC 2012). The State Council reported that these activities 'brought 330,000 hectares of new wetland under protection, restored 23,000 hectares of existing wetland, established four new wetlands of international importance and developed 68 pilot national wetland parks' (Information Office of the State Council, PRC 2012).

Marine ecosystems are also highly susceptible to climate-related impacts, and China's State Oceanic Administration (SOA) has programs to promote conservation in marine ecosystems. SOA has also introduced marine disaster risk assessment and revised marine zoning regulations, while supporting local costal governments with island restoration projects (Information Office of the State Council, PRC 2012). A 2012 fund created to promote island protection included 200 million yuan from the central government to support 15 local island protection projects (Information Office of the State Council, PRC 2012).

5.3.3 Managing Water Resources

Water resources scarcity is one of China's biggest challenges as it continues to develop, and climate change is likely to put additional strains on China's water supply. Government climate plans call for enhanced water resources management, as well as river rehabilitation, and enhanced and more unified management of water resource basins. For example, the State Council's *Opinions on Implementing the Strictest Water Resources Management System* guides the country's water resources management with a strict water resources control system consisting of three 'red lines': (1) the control of the development and use of water resources; (2) the control of water use efficiency; and (3) the restriction of pollutants in functional water resource areas (Information Office of the State Council, PRC 2012). The government reports progress being made in implementing the permit system that regulates water resource withdrawal, as well as the reimbursable usage system controlling water resources. Other key policies regulating water resources in China that are due for implementation during the *12th Five-Year Plan* period (2011–15) by a variety of government agencies include the *National Rural Drinking Water Safety Project*, the *National Water Resources Development Plan*, the *National Utilization and Protection of Underground Water* and the *Notice on Further Promoting the Construction of Water-saving Enterprises* (Information Office of the State Council, PRC 2012).

While there are clearly numerous policies and regulations in place to manage China's water resources, it remains to be seen whether they will be implemented effectively. The competing demands for water resources in China rarely put conservation as a high priority. For example, the south–north water diversion project is a massive infrastructure project that aims to link the Yangtze, Yellow and Huaihe and Haihe Rivers and redistribute water resources around the country. This project is also mentioned as a key element of the government's climate adaptation plans, even though these rivers are seeing decreasing flows due to the diversion of water resources, as well as extensive hydroelectric developments along the rivers. In addition, there are real concerns with the scale of the south–north water transfer project, particularly because the canal risks seawater infiltration along the unstable coastal areas it crosses through, especially if sea levels rise (Piao et al. 2010). Flooding along these major rivers is also a big challenge in China that will only increase with sea level rise, and these large infrastructure projects aim to reduce flood risk. A recent World Bank project targets flood management along the Huai River Basin and aims to protect over 6.6 million people from flooding along the river (World Bank 2010a). The Chinese government reports that in 2011, these policies allowed the country to 'overcome the severe autumn floods on the Yangtze, Lancang and Yellow Rivers,' and to 'successfully deal with the impact of seven typhoons and tropical storms in 2011'; the government also reported a death toll from flooding that was 'the lowest since the foundation of the People's Republic of China in 1949' (Information Office of the State Council, PRC 2012).

China's Ministry of Health has issued several regulations that aim to ensure the supply of safe drinking water, including the *Guidance on Strengthening the Supervision and Monitoring on Drinking Water Quality*, the *National Urban Drinking Water Safety Protection Plan (2011–2020)*, the *Notice on Further Strengthening the Supervision and Monitoring of Drinking Water Quality* and the *2012 National Drinking Water Supervision and Monitoring Work Plan*. In addition, the Ministry of Health established a national drinking water standard and a water quality monitoring system for drinking water with outposts in each province, accompanied by a 220 million yuan investment to support provincial-level implementation and enforcement (Information Office of the State Council, PRC 2012).

Other goals mentioned in China's climate plans, including a desire to change people's way of considering water resources as 'inexhaustible,' may be harder to achieve in the absence of pricing systems that create disincentives for wasting water. In addition, water pollution is so widespread that tackling water pollution would go a long way toward addressing water scarcity and safety issues in China.

5.3.4   Increasing the Resilience of Coastal Zones

China's 2007 *Climate Change Program* called for the development of an integrated coastal zone management system, as well as the strengthening of strategies to address sea level rise. The vulnerability of China's coastal cities has led to many municipal or city-level government initiatives to strengthen infrastructure. In particular, the Shanghai municipal government has been extremely proactive in protecting the resilience of this vulnerable coastal city to climate-related impacts.

Since 1949, Shanghai has invested approximately 50 billion yuan in flood prevention measures, with 'multiple lines of defense' (including sea walls, river embankments, flood discharge areas and urban drainage systems) now in place to help protect against flooding and storm surges (Liu 2012). Shanghai's 523 kilometers of sea walls act as a first line of defense against floods, and can withstand 'a once-in-200-years high tide level and cope with gales up to 133 kilometers per hour' (Liu 2012). In addition, Shanghai can withstand a once-in-1000-years high tide level due to the Huangpu River levee and related urban flood control projects (Shanghai Daily 2012).

Shanghai's second line of defense is in the form of walls and levees along the Huangpu River, which range from 5.86 to 6.9 meters tall and are designed to exceed the height of a severe flood experienced in 1997 (reported to reach 5.72 meters) (Shanghai Daily 2012). Shanghai's third and fourth lines of defense are in the form of flood discharge systems such as controlled water gates and urban drainage systems. Due to these defenses, Shanghai government leaders have rejected claims that it is the city that is most vulnerable to serious flooding of all the other coastal metropolises around the world (Balica et al. 2012). Other agencies within the Shanghai municipal government are actively putting in place flood control measures to protect the city. For example, the Shanghai Water Authority is reportedly building a US$3.96 billion drain system to deal with heavier floods, creating an underground reservoir to store large amounts of water (Balica et al. 2012). The drainage systems of major Chinese cities including Beijing, Shanghai and Guangzhou are reportedly based on former Soviet Union designs where a drier climate made slimmer pipes sufficient (Wang 2012). Floods in Beijing made the national news in July 2012 when streets, subways and homes flooded and a reported 77 people died (Wang 2012). Flooding also interferes with development when construction projects are delayed. During the summer of 2012, there were an estimated 8000 active construction sites across the city, including 77 subway stations and a new Disneyland theme park, all of which risk damage from flooding during construction (Liu 2012).

Another real challenge facing Shanghai is that it is sinking. The city, at the mouth of the Yangze River, has soils that are characterized as fine silty clay (Xu et al. 2003). These soils are soft, and are reportedly being compressed by the city's high-rise buildings. However, most of Shanghai's land subsidence (an estimated 70 percent) is due to groundwater being overdrawn, with the weight of skyscrapers expected to account for the other 30 percent (Liu 2012; Springer 2012). It is estimated that since 1921, Shanghai has descended more than 6 feet (Springer 2012).

5.3.5 Improving Scientific Capacity to Predict and Monitor Climate Impacts

The Chinese government has invested in improving and modernizing its climate monitoring capacity over the past several years. The Chinese Meteorological Administration has been involved in developing inventories of China's greenhouse gas emissions, as well as studies assessing climate change impacts across the country (Information Office of the State Council, PRC 2012). The SOA has begun carrying out research to better predict weather events such as El Niño and La Niña and understand their impacts on climate change, and it is monitoring the marine carbon cycle and air–sea CO_2 exchange in Chinese waters (Information Office of the State Council, People's Republic of China 2012). Increasing the ability of China's own scientists and climate science-related agencies will be extremely important for the country's ability to continue to utilize the most advanced models and tools for predicting the likely impacts of climate change and the implications for China.

5.3.6 Disaster Reporting for Extreme Weather and for Infectious Disease

Within the government institutions responsible for both extreme weather events and public health events, new resources are being devoted to disaster preparedness and early reporting systems that aim to warn the public about events that already occur but are expected to increase in severity or frequency with climate change, such as severe storms or disease outbreaks. Government entities (such as the National Disaster Reduction Committee) are being established to improve national monitoring and early warning mechanisms for natural disasters, particularly emphasizing early warning capabilities for extreme weather and climate events such as typhoons and floods (Information Office of the State Council, PRC 2012).

The reporting systems for infectious and communicable diseases are also reportedly being improved. New government programs aim to prevent and control the vector-borne diseases that are particularly susceptible to climate change, such as dengue fever and hand, foot and mouth disease. By the end of 2011, all provincial-level disease prevention and control institutions had been integrated into a national direct network reporting system to streamline surveillance, reporting, and prevention (Information Office of the State Council, PRC 2012).

5.4   CONCLUSIONS

While China is already taking significant steps to mitigate its CO_2 emissions (Lewis 2011b, 2013) the central and local governments have only just begun to implement the adaptation measures that will be crucial to China's ability to endure the expected climate-related impacts in the coming years. Experience to date has demonstrated that climate change adaptation cannot be a piecemeal strategy. Building societal resilience to climate change will require an integration of preventative strategies into all new development activities, and a reassessment of all existing risks and vulnerabilities in order to avoid the potentially devastating impacts of climate change.

Hurricane Sandy, which struck the eastern coast of the United States in October 2012, demonstrated that even a highly modern and prosperous urban center like New York can be extremely vulnerable to severe weather events. Since climate change is expected to exacerbate many of China's existing vulnerabilities, including water scarcity, increasing desertification and flooding-prone low-lying coastal cities, taking action to address these problems has an immediate benefit no matter how severe the added impact of climate change. Acting now to improve climate resilience across the country is a worthwhile and low-risk strategy that will address many of China's existing climate-related vulnerabilities, while preparing the country to deal with projected future impacts from anthropogenic climate change.

Shanghai, China's economic epicenter, is perhaps not surprisingly the site of an aggressive program to increase the city's resilience. Led by the local Shanghai municipal government, the city's actions may serve as a model for coastal China. While many of these actions are being taken in response to extreme weather events and are not directly linked to climate change adaptation strategies, they nonetheless begin to put Shanghai on the right track to being more resilient to future climate-related impacts. A vast range of vulnerabilities across the country, however, will require a diverse set of adaptation responses. China's *National Climate Program* provides a good basis for adaptation policies and planning, but the task is vast, and rapid economic development predominantly trumps thoughtful, climate-resilient development, particularly when resources are scarce.

Increased attention to adaption strategies should be a priority topic for international cooperation with China on climate change, supplementing the broad range of existing cooperation initiatives that focus almost exclusively on mitigation. For example, the newly formed United States–China Climate Change Working Group (US Department of State 2013) might consider increased attention to climate adaptation, especially as the United States and China share some similar vulnerabilities and can learn from each others' experiences. Surely Shanghai would benefit from an understanding of New York's vulnerabilities in face of Hurricane Sandy, and be involved in efforts to rebuild with increased resilience. Cooperation targeting adaptation strategies may also serve to build a mutual understanding between the two largest greenhouse gas emitters about the truly daunting challenge we face in a climate-changed world, and additionally help to form a basis for expanded cooperation to mitigate our contributions to anthropogenic climate change.

ACKNOWLEDGMENTS

The author is grateful for the research assistance of Ellen Ebert and Phillip Hah.

NOTES

*   Parts of this chapter have been extracted and updated from Joanna I. Lewis (2009), 'Climate change and security: examining China's challenges in a warming world', *International Affairs*, **85** (6), 1195–213.
1.   In 2012, about 600,000 tons of rice were exported and 486,000 tons imported, and about 1000 tons of wheat were exported and 2.2 million tons imported. These are all rather small flows compared with domestic production.
2.   At the Warsaw climate negotiations in late November 2013, the Chinese government, led by the National Development and Reform Commission, released a new "national strategy to adapt to climate change" (国家适应气候变化战略). Due to the timing of the publication of this chapter this plan is not reviewed here.

REFERENCES

ANI (2010), 'China's glaciers may shrink 27% by 2050: report', 9 October, *Daily News and Analysis*, available at http://www.dnaindia.com/world/report_china-s-glaciers-may-shrink-27pct-by-2050-report_1450309 (accessed 3 October 2012).

Arnold, R.G., D.O. Carpenter, D. Kirk et al. (2007), 'Meeting report: threats to human health and environmental sustainability in the Pacific Basin', *Environmental Health Perspectives*, **115** (12), December, 1770–5.

Balica, S., N. Wright and F. van der Meulen (2012), 'A Flood Vulnerability Index for coastal cities and its use in assessing climate change impacts', *Natural Hazards*, **64** (1), 73–105.

Chen, H.P., J. Sun and X. Chen (2012), 'Future changes of drought and flood events in China under a global warming scenario', *Atmospheric and Oceanic Science Letters*, **6** (1), 8–13.

Chovanec, P. (2011), 'Closing the gap between China's coast and interior, An American Perspective from China, 13 January, available at http://chovanec.wordpress.com/2011/01/13/closing-the-gap-between-chinas-coast-and-interior/ (accessed 12 February 2013).

Economy, E. (2007), 'China vs. Earth', *The Nation*, available at http://www.thenation.com/doc/20070507/economy (accessed 31 December 2010).

Energy Information Administration (2011), *International Energy Outlook*, Washington, DC: US Department of Energy.

Fan, D. and C. Li (2006), 'Complexities of China's coast in response to climate change', *Advances in Climate Change Research*, **2** (1), 54–8.

FAO (2012), The Food and Agricultural Organization of the United Nations Database, FAO, available at http://faostat.fao.org (accessed 2 October 2012).

Feng, W., M. Zhong and H.Z. Xu (2012), 'Sea level variations in the South China Sea inferred from satellite gravity, altimetry, and oceanographic data', *Science China Earth Sciences*, **55** (10), 1 October, 1696–701.

Government of China (2006), *National Assessment Report on Climate Change*, Beijing: Government of China.

Information Office of the State Council, PRC (2012), *China's Policies and Actions for Addressing Climate Change*, available at http://www.china.org.cn/government/whitepaper/node_7172407.htm (accessed 13 March 2013).

IPCC (2013), 'Summary for Policymakers', in T.F. Stocker, D. Qin, G.-K. Plattner, M. Tignor, S. K. Allen, J. Boschung, A. Nauels, Y. Xia, V. Bex and P.M. Midgley (eds), *Climate Change 2013: The Physical Science Basis. Contribution of Working Group I to the Fifth Assessment Report of the Intergovernmental Panel on Climate Change*, Cambridge, UK and New York, NY: Cambridge University Press.

Keeling, C.D., S.J. Walker, S.C. Piper and A.F. Bollenbacher (2013), 'Atmospheric CO_2 concentrations (ppm) derived from in situ air measurements at Mauna Loa, Observatory, Hawaii, Scripps Institute of Oceanography CO_2 program', La Jolla, California, 7 March, available at http://scrippsco2.ucsd.edu/data/in_situ_co2/monthly_mlo.csv (accessed 12 March 2013).

Kim, E.-S., D.K Park, X. Zhao et al. (2006), 'Sustainable management of grassland ecosystems for controlling Asian dusts and desertification in Asian continent and a suggestion of eco-village study in China', *Ecological Research*, **21** (6), November, 907–11.

Le Quéré, C., R.J. Andres, T. Boden et al. (2012), 'The global carbon budget 1959–2011', *Earth System Science Data Discussions*, **5** (2), 2 December, 1107–57.

Lewis, J.I. (2009), 'Climate change and security: examining China's challenges in a warming world', *International Affairs*, **85** (6), 1195–213.

Lewis, J.I. (2011a), 'China', in D. Moran (ed.), *Climate Change and National Security: A Country-level Analysis*, Washington, DC: Georgetown University Press, pp. 9–26.

Lewis, J.I. (2011b), 'Energy and climate goals in China's 12th Five-Year Plan', Pew Center on Global Climate Change, available at http://www.pewclimate.org/docUploads/energy-climate-goals-china-twelfth-five-year-plan.pdf (accessed 1 August 2013).

Lewis, J.I. (2013), *Green Innovation in China: China's Wind Power Industry and the Global Transition to a Low-carbon Economy*, New York, NY: Columbia University Press.

Lewis, J.I. and K.S. Gallagher (2010), 'Energy and environment in China: achievements and enduring challenges', in S. Van Deveer and R. Axelrod (eds), *The Global Environment: Institutions, Law and Policy*, 3rd edn, Washington, DC: CQ Press, pp. 259–84.

Li, C., D. Fan, B. Deng and D.J. Wang (2001), 'Some problems of vulnerability assessment in the coastal zone of China', in Proceedings of APN/SURVAS/LOICZ Joint Conference on Coastal Impacts of Climate Change and Adaptation in the Asia-Pacific Region, *Global Change and Asian Pacific Coasts*, Kobe, Japan, pp. 49–56.

Lin, E., Y. Xu, S. Wu, H. Hu and S. Ma (2007), 'Synopsis of China National Climate Change Assessment Report (II): climate change impacts and adaptation', *Advances in Climate Change Research*, **3**, 6–11.

Liu, X. (2012), 'A government briefing on this year's preparation against floods', 27 July, available at http://en.shio.gov.cn/presscon/2012/06/27/1151902.html (accessed 2 October 2012).

Liu, C., Y. Wang, M. Shao and X. Hou (eds) (2012), *Water Science & Technology in China: A Roadmap to 2050*, New York and Beijing: Springer and Science Press.

Luo, L.J. (2010), 'Breeding for water-saving and drought-resistance rice (WDR) in China', *Journal of Experimental Botany*, **61** (13), 1 August, 3509–17.

McGranahan, G., D. Balk and B. Anderson (2007), 'The rising tide: assessing the risks of climate change and human settlements in low elevation coastal zones', *Environment and Urbanization*, **19** (1), 1 April, 17–37.

McMichael, A.J., D. Campbell-Lendrum, S. Kovats et al. (2004), 'Global climate change', in M. Ezzati, A.D. Lopez, A. Rodgers and C.J.L. Murray (eds), *Comparative Quantification of Health Risks: Global and Regional Burden of Disease Attributable to Selected Major Risk Factors*, Geneva: World Health Organization, Chapter 20, available at http://www.who.int/healthinfo/global_burden_disease/cra/en/ (accessed 2 October 2012).

National Development and Reform Commission (2007), *National Climate Change Program*, available at http://www.china.org.cn/english/environment/213624.htm#16 (acessed 9 October 2012).

NBS (National Bureau of Statistics) (2011), *China Statistical Yearbook*, Beijing: China Statistics Press.

Nicholls, R.J., S. Hanson, C. Herweihjer and N. Patmore (2007), *Ranking of the World's Cities Most Exposed to Coastal Flooding Today and in the Future*,

Paris: OECD, available at https://support.rms.com/publications/OECD_Cities_ Coastal_Flooding.pdf (accessed 12 March 2013).

Peng, S., J. Huang, J.E. Sheehy et al. (2004), 'Rice yields decline with higher night temperature from global warming', *Proceedings of the National Academy of Sciences of the United States of America*, **101** (27), 6 July, 9971–5.

People's Daily Online (2008), 'Desertification causes yearly loss of 54 billion yuan in China', *People's Daily Online*, 26 November, available at http://english.people.com.cn/90001/90776/6541192.html (accessed 9 May 2013).

Piao, S., P. Ciais, Y. Huang et al. (2010), 'The impacts of climate change on water resources and agriculture in China', *Nature*, **467** (7311), 2 September, 43–51.

Podesta, J., P. Ogden, K. Campbell et al. (2007), 'Security implications of climate scenario 1: expected climate change over next 30 years', in K.M Campbell, J. Gulledge, J.R. McNeill et al. (eds), *The Age of Consequences: The Foreign Policy and National Security Implications of Global Climate Change*, Washington, DC: Center for Strategic and International Studies, pp.55–60.

Poole, J. and R. Ruitenberg (2011), 'China, largest wheat grower, facing threat of drought', *Bloomberg*, 8 February, available at http://www.bloomberg.com/news/2011-02-08/drought-in-north-china-plain-threatens-wheat-production-fao-report-says.html (accessed 12 March 2013).

Secretariat of the China National Committee (2000), *China National Report on the Implementation of United Nations Convention to Combat Desertification and National Action Programme to Combat Desertification*, Beijing: CCICCD.

Shanghai Daily (2012), 'Shanghai official rejects flooding claims', 23 August, available at http://www.china.org.cn/environment/2012-08/23/content_26315277.htm (accessed 2 October 2012).

Solomon, S., D. Qin, M. Manning et al. (eds) (2007), 'Summary for policymakers', in *Climate Change 2007: The Physical Science Basis. Contribution of Working Group I to the Fourth Assessment Report of the Intergovernmental Panel on Climate Change*, Cambridge and New York: Cambridge University Press, pp.1–17.

Springer, K. (2012), 'Shanghai is sinking: how building up is bringing it down', *Time*, 21 May, available at http://science.time.com/2012/05/21/soaring-to-sinking-how-building-up-is-bringing-shanghai-down/ (accessed 13 March 2013).

US Department of Agriculture (2009), *World Agricultural Supply and Demand Estimates*, United States Department of Agriculture, available at http://www.usda.gov/oce/commodity/wasde/latest.pdf (accessed 2 October 2012).

US Department of State (2013), *Joint U.S.-China Statement on Climate Change*, Unites States Department of State, 13 April, available at http://www.state.gov/r/pa/prs/ps/2013/04/207465.htm (accessed 9 May 2013).

Wang, Q. (2013), 'China's sea level continues to rise', *China Daily*, 27 February, available at http://www.chinadaily.com.cn/cndy/2013-02/27/content_16258964.htm (accessed 12 March 2013).

Wang, Y. (2012), 'City to build $3.96 billion drain system to deal with heavier rains', *Global Times*, 4 September, available at http://www.globaltimes.cn/content/731091.shtml (accessed 2 October 2012).

Wang, S. and Z. Zhang (2011), 'Effects of climate change on water resources in China', *Climate Resources*, **47**, 77–82.

Wang, X., F. Chen, E. Hasi and J. Li (2008), 'Desertification in China: an assessment', *Earth-Science Reviews*, **88** (3–4), June, 188–206.

Wang, X., Y. Yang, Z. Dong and C. Zhang (2009), 'Responses of dune activity and desertification in China to global warming in the twenty-first century', *Global and Planetary Change*, **67** (3–4), June, 167–85.

Wang, J., J. Huang and S. Rozelle (2010), 'Climate change and China's agricultural sector: an overview of impacts, adaptation and mitigation', Issue Brief No. 5, International Center for Trade and Sustainable Development (ICTSD), Geneva.

Wang, J., W. Gao, S. Xu and L. Yu (2012), 'Evaluation of the combined risk of sea level rise, land subsidence, and storm surges on the coastal areas of Shanghai, China', *Climatic Change*, **115**, 537–58.

Whiteman, H. (2012), 'China doubles Beijing flood death toll', CNN, available at http://articles.cnn.com/2012–07–26/asia/world_asia_china-beijing-flood_1_death-toll-flood-control-flood-risk (accessed 2 October 2012).

WICCI: Human Health Working Group (2009), *Wisconsin Initiative on Climate Change Impacts*, available at http://www.wicci.wisc.edu/human-health-working-group.php (accessed 3 October 2012).

World Bank (2010a), *Huai River Basin Flood Management and Drainage Improvement*, Washington, DC: World Bank, available at http://www.worldbank.org/projects/P098078/huai-river-basin-flood-management-drainage-improvement?lang=en (accessed 2 Ocober 2012).

World Bank (2010b), *Cities and Climate Change: An Urgent Agenda*, Washington, DC: World Bank, available at http://climatechange.worldbank.org/content/new-report-sees-cities-central-climate-action (accessed 3 October 2012).

World Bank (2012a), *Mainstreaming Climate Change Adaptation in Irrigated Agriculture Project*, Washington, DC: World Bank, available at http://www.world-bank.org/projects/P105229/mainstreaming-climate-change-adaptation-irrigated-agriculture-project?lang=en.

World Bank (2012b), *Sustainable Low-carbon City Development in China*, Policy Note 67226, Washington, DC: World Bank, available at http://documents.worldbank.org/curated/en/2012/02/15879709/sustainable-low-carbon-city-development-china (accessed 2 October 2012).

World Economic Forum (2008), *Africa @ Risk*, available at http://www.weforum.org/pdf/Africa2008/Africa_RiskReport_08.pdf (accessed 2 October 2012).

World Health Organization (2003), *Climate Change and Human Health: Risks and Responses*, Geneva: World Health Organization.

Wu, J. (2008), 'Changing diet offers more food for thought', *China Daily*, 24 June, available at http://www.chinadaily.com.cn/china/2008–06/24/content_6790223.htm (accessed 2 October 2012).

Wu, Y. and R. Yu (2013), 'Shanghai sees less trade in 2012', *China Daily*, 22 January, available at http://www.chinadaily.com.cn/cndy/2013–01/22/content_16150880.htm (accessed 9 May 2013).

Xinhua (2013), '4.97m affected by lingering drought in SW China', *China Daily*, 9 March, available at http://www.chinadaily.com.cn/china/2013–03/09/content_16294420.htm.

Xinhua News Agency (2008), 'Guangdong reports 20% growth in foreign trade in 2007', 11 January, available at http://english.sina.com/business/1/2008/0111/141171.html (accessed 2 October 2012).

Xinhua News Agency (2011), 'National Assessment Report on Climate Change: China's sea level will continue to rise in future', 16 November, available at http://news.xinhuanet.com/tech/2011–11/16/c_122286961.htm.

Xu, Y., D. Sun, J. Sun, D. Fu and P. Dong (2003), 'Soil disturbance of Shanghai silty clay during EPB tunneling', *Tunnelling and Underground Space Technology*, **18** (5), November, 537–45.

Yu, Q. (2007), 'Special Representative for climate change negotiations of the Ministry of Foreign Affairs Yu Qingtai receives interview of the media', available at http://www.chinaembassy.org.in/ (accessed 3 January 2011).

6. Adaptation to climate change in Bangladesh: migration, the missing link

Tasneem Siddiqui and Motasim Billah

Traditional climate change literature in Bangladesh has generally treated migration with fear (Walsham 2010). Through initiation of national and sectoral policies and community-level adaptation programmes, successive governments of Bangladesh, civil society bodies and development partners have focused on creating local-level opportunities of adaptation and reducing the scope of migration of affected people. They have mostly treated migration as a consequence of failure to adapt. This chapter highlights that a section of people of the areas concerned have used migration as one of the strategies either to diversify their income sources or reduce risks to their life triggered by different sources including climate stresses. By drawing on evidence from current experiences of people and communities at risk this chapter argues that different types of migration help build resilience for both those who move and those who choose not to. Moreover, this chapter suggests increasing the scope for members of affected families to exploit the existing opportunities of internal migration as well as short-term international migration as one of many adaptation strategies.

The chapter is divided into seven sections. Section 6.1 deals with theoretical issues. Section 6.2 sets the country context. Section 6.3 reviews existing literature on the impact of climate change on the lives and livelihoods of affected people and their experience of migration. Section 6.4 looks at laws and policies that have been initiated to systematically respond to reduce the harmful effects of climate change. Section 6.5 presents adaptation programmes currently being implemented at community levels. Section 6.6 provides evidence from primary research of the use of migration as an adaptation tool by people who are facing shocks of different climatic events. Section 6.7 draws major conclusions and makes suggestions for policy reorientation.

The chapter is based on both primary and secondary information. Primary information was gathered in three districts of Bangladesh,

Satkhira, Nawabganj and Munshiganj. Four villages from each of these districts were selected on the basis of their exposure to different types of environmental stresses. These include coastal cyclone and storm surge, saline intrusion, drought, flood and riverbank erosion. Primary research collected two types of data: history of the villages and migration experiences of the villagers. Key informants of 12 villages were interviewed and focus group meetings were conducted with a cross-section of villagers. In each of the focus group discussions 30 to 40 villagers participated.

6.1 THEORETICAL ISSUES

Although for quite some time displacement of a large number of people has been identified as an important consequence of climate change, only recently has empirical evidence-based research been undertaken to understand the relationship between migration and climate change. The 16th session of United Nations Framework Convention on Climate Change (UNFCCC), held in 2010, has called for a better understanding of the issue (UNFCCC 2011). Theories of migration underscore the complexity of identifying one-to-one causal relationship with migration and its determinants, as migration is multi-causal. Along with traditional push-pull models, historical structuralism, human capital or rational choice theories, theories of new economics have broadened understanding of the causes of migration by adding household decision-making models, migration as risk minimisation or an insurance strategy. Social network theory helps to explain the role of intermediaries in increasing access to migration for a large group of people. Over the years all these frameworks have contributed towards development of a robust theoretical understanding of the interaction of multiple causes of migration.

Most theoretical frameworks incorporate environmental change as a structural or influencing factor for migration of individuals or communities. However, predictions made in the mainstream climate change literature about the movement of millions of people from different parts of the world due to environmental stresses and shocks are not based on any such analysis. Migration researchers initially drew strong links between past climatic events and migration (Hugo 1996). They contributed in providing terminological clarity on how to define people who move in response to environmental or climate-related stresses. As refugee situations entail fear of persecution, they introduced the term 'environmental migrants' instead of 'climate refugees' or 'environmental refugees'. The International Organization for Migration (IOM) came up with a working definition of environmental migrants. Environmental migrants are 'those persons who for reasons of sudden or progressive changes in the environment that adversely affect their lives or living conditions are obliged to leave their habitual homes, or choose to do so, either temporarily or permanently and who move either within their country or abroad' (IOM 2008, p. 1).

Kniveton et al. (2009) demonstrated that the relationship between migration and climate change is not linear. It affects different groups of people differently. A particular environmental event may increase migration in one context; the same event in another context or at a different time can decrease migration. The deteriorating rainfall in sub-Saharan Africa caused an increased rural to urban migration (Kniveton et al. 2008), whereas the drought-affected people of Mali could not migrate to cities because of their inability to find needed finances (Foresight 2011). Arongo (2004), Massey et al. (1998) and others have shown that push-pull factors along with intervening factors operate at the level of human agency and create conditions that lead some to migrate and restrict the migration of others. Martin et al. (forthcoming a) further elaborated that push-pull and intermediary determinants of migration each have economic, social, political, demographic and environmental influences. In other words, environment is one of the many stimuli that work within push-pull and intermediary factors. For such complexity instead of using the term 'environmental migrants' Foresight (2011) used the term 'migration influenced by environmental change'. The Foresight report stressed that environmental change will influence different drivers of migration and thus play a role in effecting migration as well as trapping a section of poor people in vulnerable areas.

Adaptation is a major issue in recent climate change discourse. Although some understanding is emerging about how climate change influences migration, to what extent affected people use migration as one of their adaptation strategies has hardly been explored. Instead of viewing the migration of climate change-affected people as a threat, Siddiqui (2009, 2010a, 2011) argued for transforming migration as one of the tools for adaptation where it is feasible. At this stage, it is important to define adaptation. Adaptation is closely linked to the socio-ecological concept of resilience. Resilience is seen as 'the ability of human systems to learn from and reorganise to meet changed conditions' (Dolan and Walker 2006, p. 1319). Resilience is defined as the ability to absorb disturbance, undergo change, then self-organise, maintain the same basic framework and method of function and learn from and adapt to the disturbance (Dolan and Walker 2003).

In other words, it is the capacity of the human system to absorb disturbance. The process of absorption is the function of adaptation. McCarthy et al. (2001) viewed adaptation as the response of natural and human systems to adjust with real or expected climatic events or exploit beneficial opportunity. Grothmann and Patt (2005) showed that past adaptation experience can have a positive effect on future adaptation. Agarwal (2008) has argued that adaptation to climate change should always be at the local level.

Some experiences of migration of affected people can be theorised as functions of adaptation. For example, when people are moving out of areas exposed to further risk they are perhaps adapting to the disturbance of climate change. Short-term migration during natural disasters such as floods or resettlement at a new area when homestead land is lost due to riverbank erosion are essentially examples of adaptation. Seasonal and circular migration of a few members of affected households to reduce dependency on local resources for livelihoods or to diversify income sources can be explained as the enhancement of resilience through innovation and adaptation. The attempts of affected households to enhance income by receiving remittances from their permanent or short-term migrant family members can also be explained as the use of migration as an adaptation technique. But there is a subtle difference between coping and adaptation. Migration experiences may be treated as coping mechanisms unless people can maintain or enhance their previous economic wellbeing through migration. Systematic empirical inquiries are necessary to engage policy makers in incorporating migration into their adaptation strategies as well as mainstream development programmes. The concept of resilience allows migration to be seen as part of the co-evolution of socio-ecological systems rather than a socio-economic process disconnected from the environment (Martin et al. forthcoming b).

6.2 INTERNAL AND INTERNATIONAL MIGRATION AND THE ECONOMY

During the initial years of independence, Bangladesh was referred to as a 'bottomless basket' by some of its development partners. Over the years, through the hard work of its people, Bangladesh has experienced significant social and economic progress. Its gross domestic product (GDP) has grown 5.3 to 6.7 per cent per annum over the last ten years. With only 147,570 square kilometres of land and a population of 160 million, the country raised its per capita income to US$848 in the fiscal year 2011–12 (GoB 2013). Bangladesh's population growth rate declined from 2.9 per cent in 1974 to 1.2 per cent by 2012. People living below the national poverty line have declined from 40 per cent in 2005 to 31.5 per cent in 2010 (HIES 2010).[1] Its Human Development Index improved from 0.462 in 2005 to 0.500 in 2011 (UNDP 2012). Rahman (2010) showed that even with its current growth rate, the country will reach the status of middle-income country before 2020.

The three major contributors to this development are garment manufacturing, services and remittances sent by Bangladesh's international migrant workers. In 2012, migrants remitted US$14.17 billion. This is 44 per cent of the total remittances received by the 48 least developed countries (LDCs). This is six times more than the overseas development aid the country received and three times more than the net earnings from the garment manufacturing sector. The garment sector is mostly driven by internal migrants and remittances comprise solely of contributions from international migrants. Therefore, the role of internal and international migration to the economic development of Bangladesh can hardly be overemphasised.

Global climate changes can compromise Bangladesh's hard-earned economic and social gains. Bangladesh is extremely vulnerable to climate change. Some even argue that climate change is no longer a future threat for Bangladesh as it has already affected the country (Rahman 2010). Sea level rise, flood, drought, cyclone, saline intrusion in coastal farmlands and water logging are some of the major environmental challenges that the country faces due to global climate change. Considering the climatic characteristics of Bangladesh, the assessment of the *International Strategy for Disaster Reduction* placed Bangladesh at the top of the list of countries vulnerable to disasters when it comes to vulnerability to floods, third in respect to tsunami and sixth in respect to cyclone (Ali 2010).

6.3 EVIDENCE OF USE OF MIGRATION IN EXISTING LITERATURE

This section reviews existing literature on the impact of environmental stress on the livelihoods of affected people. It attempts to locate the experience of migration cited in the literature. It concentrates on four types of environmental stress: flood, cyclone, drought and riverbank erosion.

6.3.1 Flood

Bangladesh is deltaic country at the confluence of three rivers: *Padma, Jamuna* and *Meghna*. It has a dense network of more than 230 rivers. Although the territory of Bangladesh constitutes only 8 per cent of the Ganges drainage basin, 93 per cent of the water of this region (Ganga-Brahmaputra-Meghna Basin) flows over Bangladesh to the Bay of Bengal which makes the country naturally flood-prone.

Floods are a regular feature in Bangladesh. Generally, a quarter of the landmass is affected by floods in a typical year. The frequency of severe floods has intensified in the last 25 years. Cases of protracted water-logging following floods have also increased. In 1988, floods displaced 45 million people and in 1998, displaced 30 million people. The flood of 2007 affected almost 16 million people, destroying 85,000 houses, and damaging another 1 million. Approximately 1.2 million tons of crops were destroyed or partially damaged. The estimated damage amounted to over US$1billion (GoB 2007). The last three floods of 2005, 2006 and 2007 altogether destroyed 0.14, 0.26 and 1.5 million hectares of crop land, respectively (Rahman et al. 2007). The small farmholders suffer more due to the damage caused to their standing crops, leading to increased debt. Some migrate to a close proximity to diversify income sources. There are substantive data on flood-induced primary migration but there still remains little evidence that floods influence longer-term migration decisions. Rayhan and Grote (2007) conducted a survey covering 595 households. It showed that 28 per cent of these households had at least one migrant; 83 per cent reported that they had migrated because of lack of employment as a result of frequent floods; only 6 per cent moved to different villages. Most of these families were very poor: 89 per cent went to a nearby city[2] and the 5 per cent of the remaining 11 per cent, who mostly belonging to the richest quartile, migrated internationally (Rayhan and Grote 2007). The study did not probe into what enabled the remaining 72 per cent to stay in the inundated village. Paul (2003), however, showed that most people do not move from affected areas after floods if disaster aid programmes are implemented effectively.

6.3.2 Cyclone and Storm Surge

More than two-thirds of the 19 coastal districts in Bangladesh are directly exposed to cyclones and storm surges. Since 1970 Bangladesh has experienced a total of 26 major cyclones, 18 in the last 20 years. These cyclones affected 19 million people. Among them four major cyclones (1970, 1991, 2007 and 2009) caused massive havoc and death toll. The frequency of cyclones has increased since 1990. Recent cyclones are accompanied by high winds and storm surges up to 7 metres. Over the last six years Bangladesh has faced two high-intensity cyclones, the super Cyclone *Sidr* of 2007 and the Cyclone *Aila* of 2009.

Cyclone Sidr caused great harm to the lives and livelihoods of people. It affected 8.7 million people in 30 out of 64 districts to varying degrees; causing the deaths of over 3400 people and injuries of over 55,000 people. One million households lost their dwellings. The cyclone also badly affected people's livelihood in the coastal areas. A study by the International Labour Organization (2008) estimated that a total of 567,000 people in the 12 most affected districts lost their principal source of income either permanently or temporarily. *Sidr* damaged 54,000 shrimp enclosures and 208,000 fish ponds. The storm affected private businesses with an average asset loss of US$300 per employed person in factories, and an average of about US$100 for self-employed workers, such as van or rickshaw pullers and sewing machine operators.

Cyclone *Aila* also severely impacted the lives and livelihoods of people. It affected almost 3.9 million people, wrecked 1742 kilometres of embankments and displaced 76,478 families in Satkhira and Khulna districts (UN 2010). It badly hit *Sundarbans* the largest mangrove forest in the world. People living along the belt of *Sundarbans* pursue different types of livelihood, for example, fishing and gathering of forest resources including honey, leaves and timber. After the cyclone *Aila*, the government restricted people from pursuing their livelihoods in the *Sundarbans* to help it regenerate naturally from the destruction. The government also temporarily halted shrimp cultivation for two years in Shyamnagar sub-district of Satkhira to reduce the high level of water and soil salinity caused significantly by the sea water intrusion during *Aila*. In the post *Aila* period migration from the affected sub-districts, for example, Koyra, Paikgachha, Dacope and Batiaghata of Khulna and Shyamnagar of Satkhira district increased manifold. The migration mainly took seasonal and temporary in nature. The lack of year-round livelihood opportunities partly caused by *Aila* was assumed to be the primary reason for the growing migration flow (Martin et al. forthcoming a).

6.3.3 Riverbank Erosion

People living along the sides of rivers and *Chars* face frequent displacement due to riverbank erosion.[3] Annually it affects about 1 million people in Bangladesh (Abrar and Azad 2003). Riverbank erosion leads to decreased agricultural land and permanent loss of livelihoods particularly for those involved in agriculture. Families living alongside the vulnerable riverbank experience multiple shocks and displacements. Abrar and Azad (2003) found that each household suffered displacement in northwest region of Bangladesh on average 4.46 times due to riverbank erosion. Haque and Zaman (1989) showed that a majority of displaced people initially resettled locally within a 2.2 kilometres range whereas a small section moved a greater distance of 8 kilometres.

Skinner and Siddiqui (2005) found that under normal circumstances, 80 per cent of *Char* households from six villages of northwest Bangladesh were migrant households. Migration from *Chars* is mostly temporary, seasonal and circular in nature. Hutton and Haque (2004) found that in 1998 a total number of 5500 out of 30,000 who were residing in Sirajganj slum were erosion-affected displacees. Foresight (2011) highlights that the poorest of the poor ends up migrating to more vulnerable areas.

6.3.4 Drought

Every year northwest Bangladesh faces severe seasonal drought. Many factors are attributed to drought conditions. They include long-term changes in rainfall patterns, over-pumping of groundwater and diversion of water upstream. The intensity and duration of drought has been increasing over the years, putting major stress on the livelihoods of people in this region who are overwhelmingly agricultural farmers. In 2006 crop production was reduced by 25–30 per cent in northwestern Bangladesh due to drought (Rahman et al. 2007).

A randomised intervention project conducted in Rangpur district showed that people migrate from their places of origin during *Monga*[4] period. The findings of the study revealed that over 40 per cent of beneficiaries who received cash or credit from the project and 16 per cent of those who obtained information chose to migrate to both Dhaka as well as neighbouring towns (Chowdhury et al. 2009). The northwest region particularly Rangpur, is one of the top five districts of the origin of slum dwellers in Dhaka city; accounting for 4.6 per cent of the total slum population who went there despite its considerable distance from Dhaka (Angeles et al. 2009).

This section has shown that existing literature on flood, cyclone, drought and riverbank erosion in Bangladesh describes migration during and after environmental stresses. Floods mostly induce migration of short duration. Nonetheless, not all affected people migrate during floods. Recent cyclone and storm surges have resulted in forced displacement of affected families. Riverbank erosion also causes forced displacement but the displacees tend to move locally. During droughts, working-age men of northwest Bengal migrate to cities.

6.4   GOVERNMENT POLICIES AND LAWS ON CLIMATE ADAPTATION

In order to deal with the adverse impacts of climate change, the successive governments of Bangladesh have undertaken a series of sectoral policies, action plans and programmes. Environmental policy (1992), environmental law (1997), forest policy (1994), fisheries policy (1998), water policy (1998), new agriculture extension policy (1995), energy policy (1995), coastal zone management policy and national plans for disaster management are some of the important ones. Along with these policies, the *National Conservation Strategy* (NCS) and the *National Environment Management Action Plan* (NEMAP), the *National Adaptation Programme of Action* (NAPA) and the *Bangladesh Climate Change Strategy and Action Plan* (BCCSAP) are directly linked to achieve the goals of the above policies. The following discusses these policies, strategies and action plans to see how they have dealt with migration.

A task force was created by the Ministry of Environment and Forests in the early 1990s to prepare NEMAP and NCS. The task force included state and non-state agencies and international organisations. NEMAP (1995) proposed actions for sustainable development. It identified a list of priorities and classified them as institutional, sectoral, location-specific or long-term issues.[5] There was no mention of migration in NEMAP. While working on NEMAP the task force also prepared the environmental policy (GoB 1995).

The *National Environmental Policy* (1992) was framed even before the Earth Summit. It set the principles that the Government of Bangladesh would like to ensure in future while pursuing its broader social and economic goals. The major objectives of the environment policy are: 'To maintain ecological balance and overall development through protection and improvement of the environment; to protect the country against natural disasters; to identify and regulate activities which pollute and degrade the environment; to ensure environmentally sound development in all sectors and to ensure sustainable, long term and environmentally sound use of all national resources' (Aminuzzaman 2010, p. 2). This policy also did not deal with migration.

To reinforce the environment policy, in 1997 the Environmental Conservation Act (ECA) was passed. It was subsequently amended in 2000. The Act targets conservation and the upgrading of environmental standards with emphasis on control and mitigation of environmental pollution.

NAPA was prepared in 2005 and submitted to the United Nations Framework Convention on Climate Change (UNFCCC). NAPA covered immediate and urgent adaptation needs. It included 15 projects. Among these projects, only the 'Community Based Adaptation to Climate Change through Coastal Afforestation' project secured funding from the LDC Fund. NAPA viewed migration as an undesirable outcome of climate change and associated it with an increased incidence of crime. It proposed adaptation actions as a method to reduce the scope of unwanted social consequences of mass migration to cities. In 2009 NAPA was updated to include wider adaptation requirements and identified 45 adaptation measures of which 18 have been prioritised. It refrained however from presenting migration as a social threat.

BCCSAP was first drafted in 2008 and finalised and adopted in 2009. It is based on NAPA. It formulates a strategy and plan of action to address the adverse effects of climate change. BCCSAP emphasises institutional and human capacity development. It focuses on awareness raising and disaster preparedness. It also underscores the need to have research and data management in place. It is built on six pillars: 'food security, social protection and health, comprehensive disaster management, infrastructure, research and knowledge management, mitigation and low carbon development and capacity building and institutional strengthening' (GoB 2009, p. 3). Currently, many types of programmes are being implemented in Bangladesh under the six themes.

The 2008 draft of BCCSAP presented population movement as a challenge in different sections. It cited that 6–8 million people could be displaced, but refrained from linking them with social problems or crime. It treated rural urban migration of climate change-affected people as a problem of unplanned urbanisation. It did not have any policy direction, however, on how to treat migration in the overall adaptation action plan. The revised 2009 BCCSAP increased the figure of potentially displaced persons to 20 million. It suggested resettlement of these people through international migration. Development of human resources in climate change-affected areas is suggested to make the population competitive in the global market. Theme four, on research and knowledge management, for the first time included the migration issue. It identified the monitoring of climate change-related internal and external migration and rehabilitation as one of its tasks.

In order to implement the action plan, the government developed a four-tier institutional arrangement. The National Environment Committee is the highest body, headed by the prime minister. This high-level committee provides strategic guidance and oversight to BCCSAP. The National Steering Committee on Climate Change constitutes the second tier and is headed by the Minister of Environment and Forests. This committee is in charge of the overall coordination and facilitation of the action plan implementation. The Climate Change Unit, which is at tier 3, is also situated in the Ministry of Environment and Forests. Its function is to manage different programmes. Tier four is constituted by focal points on climate change in different ministries. These ministries plan and implement activities that fall under their purview. The concerned ministries and departments are the Ministry of Agriculture, Ministry of Fisheries and Livestock, Ministry of Water Resources, Ministry of Energy, Ministry of Health and Family Welfare, Ministry of Education, Ministry of Housing and Public Works, Planning Commission, Department of Forest and Department of Environment. One can see that the Ministry of Labour and Employment and the Ministry of Expatriates' Welfare and Overseas Employment, which deal with internal and international migration, are not part of the process. This means that BCCSAP is being implemented without their knowledge and input on migration issues.

This section has shown that the Government of Bangladesh has initiated different policies and laws to protect and conserve the environment. Most of the earlier policies are devoid of any discussion on migration. Only recently have government documents started to mention migration. Unfortunately, these documents present migration as a threat. The government's BCCSAP for the first time mentioned the need to understand the extent of migration from areas that are affected by adverse climatic events.

6.5 CIVIL SOCIETY PARTICIPATION IN COMMUNITY-BASED ADAPTATION

The civil society of Bangladesh along with its international counterparts are actively involved in global campaigns on the impact of climate change on the ecosystem of Bangladesh and the lives and livelihoods of vulnerable groups who are already affected or could be affected in future due to climate change or environmental stress. The Bangladesh Centre for Advanced Studies (BCAS), the International Union for Conservation of Nature (IUCN) Bangladesh, CARE Bangladesh, Practical Action Bangladesh, Plan Bangladesh, OXFAM Bangladesh, Action Aid Bangladesh, the Bangladesh Rural Advancement Committee (BRAC), Shushilan, Rupantor and Grameen Bahumukhi Unnayan Sangstha (GIBUS) are some of the leading international, national and local-level organisations that are deeply involved with climate change issues.

These organisations have undertaken a large number of community-based adaptation programmes in different vulnerable locations of Bangladesh. Recently OXFAM (Sterrett 2011) published a review of adaptation practices in South Asia. Among the five countries under review, Bangladesh had the largest number of programmes. Participatory research, gathering and use of indigenous knowledge and technology, awareness raising, training, livelihood support for vulnerable groups and advocacy are some of the important approaches used by almost all the

non-governmental organisations (NGOs). The 3rd international conference of BCAS on community-based adaptation to climate change in 2009
identified five major areas where NGOs have been experimenting with
community-based adaptation. These are the vulnerability assessment exercises, practising climate-resilient agriculture, use of indigenous knowledge
in disaster risk reduction, education and awareness and local-level capacity building. Some of the programmes are assessed below to see how they
address the issue of migration (BCAS 2009).

Since 2002 CARE Bangladesh has been implementing a programme entitled 'Reducing Vulnerability to Climate Change (RVCC)' in Jessore, Narail,
Gopalganj, Satkhira, Khulna and Bagerhat in southwest Bangladesh.
This project aims to promote sustainable development and local capacity
with the help of 16 partner NGOs. It targets increasing agricultural food
production, alternative livelihoods, increased availability of food, access
to safe water, safe housing, improved health and personal safety, and
increased access to common property resources (Sterrett 2011, p. 30). The
programme also targets increasing the capacity of local partners to collect
and disseminate information related to climate change through awareness
campaigns and advocacy. A number of important achievements of the
projects have been identified by an OXFAM review (Sterrett 2011). These
are the development of mangrove nurseries, grass cultivation at riverbanks,
pond, pisciculture, cage agriculture, and the production of salt-resilient,
drought-resilient and flood-resilient crop varieties. Migration issues did
not surface in the CARE projects.

A national NGO Uttoron has been working to reduce the effect of
flood, waterlogging and sea level rise. In a one-year period, it raised 650
hectares of land using traditional tidal river management techniques on
the *Hari* River Basin in southwest Bangladesh. In *Bhayna Beel* area, this
intervention has increased the flow of water in the river and decreased the
scope of waterlogging (Sterrett 2011).

Action Aid Bangladesh, an international NGO, worked on climate
change adaptation programmes in 12 villages of Sirajgong, Naogaon
and Patuakhali districts during 2008–10. The programmes dealt with the
adverse impacts of drought, flood, cyclones, sea level rise and saline intrusion by experimenting with different adaptation options. The programme
used ring wells, tube wells, climate-friendly latrines and pond excavation for
improved water and sanitation. It distributed handloom machines, sheep,
goats and ducks, vegetable and fruit tree seedlings to support people's
livelihoods in difficult conditions. It also formed climate-resilient cluster
villages and imparted training to local government officials, journalists and
local politicians on climate change issues. Action Aid conducted research
on levels of migration from three climate change-affected areas.

IUCN Bangladesh and Practical Action, Dhaka jointly undertook a programme entitled 'Participatory Vulnerability Assessment Exercises' (VAE) between January 2008 and September 2009. The project used VAE as a tool for analysing effective links between climate change and local knowledge and practices in 21 *Unions*[6] of Sadar and Subarnachar sub-districts of the Noakhali district in order to develop a disaster vulnerability map of those areas. The exercise ensured the participation of local people of various professions and age groups including farmers, day labourers, fishermen, women and children to understand current and future vulnerabilities. It also tried to take into account the existing coping mechanisms of local people. The findings indicate that waterlogging is the main problem of the areas and is aggravated by drainage congestion. Further, the long spells of inundation and increased soil salinity affect cropping patterns. Unfortunately, none of these exercises enquired about the level of use of migration as a coping mechanism or, for that matter, as an adaptation tool.

In the coastal district of Noakhali, the IUCN has been implementing the 'Promotion of Adaptation to Climate Change and Climate Variability Project'. One of the programmes under this project raised and reinforced homesteads to make them more resilient to cyclones. The project also helped local people plant trees along the edges of their homesteads to reduce wind impacts.

In 2008, the IUCN and the Bangladesh Char Development and Settlement Programme (CDSP) implemented an awareness-raising programme in four Sub-districts of Noakhali. The programme aimed to make people particularly students and teachers aware of issues and concerns related to climate change and climate variability. The project educated the youth of the coastal communities about climate change and attempted to increase their capacity to face climate vulnerabilities. A scrutiny of their awareness campaign content shows that they did not contain any information about the scope for diversifying family income by the participation of one or a few household members in internal and international labour markets.

This section has shown that different organisations have undertaken programmes of adaptation. The main goal of all these programmes is to help local people develop their capacity to continue their lives and livelihoods in the changed situations arising out of climatic stress. Nonetheless, these projects hardly deal with the migration experience of local communities. Some of these projects are particularly geared to gather local knowledge and conventional wisdom with respect to waterlogging, flood or drought. Unfortunately, these knowledge-generation efforts do not try to learn how local people are using migration as an adaptation strategy. This suggests that all these actors have a preconceived notion that the migration of people implies their adaptation programmes are not functioning well. A reasoned discussion for dealing with migration is also absent.

6.6   USE OF MIGRATION AS AN ADAPTATION TOOL

This section presents the empirical findings on the use of migration by poor people affected by environmental stresses and shocks. Satkhira, Nawabganj and Munshiganj are three typical districts that experience different types of environmental stresses and shocks. Satkhira is one of the hard-hit coastal districts of Bangladesh that experiences periodic cyclone and storm surges resulting in displacement, saline water intrusion, water-logging and loss of livelihood. Nawabganj experiences seasonal drought and flood. Munshiganj characterises a case of riverbank erosion and flood. Four villages from each district were studied.

6.6.1   Satkhira

Study of four villages, namely Gabura, Maddhayam Khailashbunia, Padmapukur and Khutikata of Satkhira district demonstrates that in the past, the livelihoods of the villagers were based on agriculture and agriculture-related professions and resources generated from the *Sundarbans*. The poorer section of the villagers used to migrate to other districts seasonally during sowing and harvest time. During the 1990s, shrimp cultivation became popular in the locality and the importance of farm-based agriculture diminished. Shrimp cultivation was less labour-intensive compared to paddy cultivation. Various restrictions also started to be imposed on gathering resources from the *Sundarbans*. By 2012 major shifts in profession became visible compared to the 1970s. Previously, a household's earning members pursued multiple professions such as subsistence farmer, sharecropper, agricultural labourer, boatman, forest wood cutter, honey and leaf collector, potter and smith. Under the changed circumstances, the poor people of the village work as labourers in shrimp farms, day labourers, petty traders, shopkeepers and seasonal migrants. Employment opportunities as seasonal migrants have also become diversified. Along with seasonal agriculture work they have been absorbed in brickfields, shrimp fry collection, paddy processing mills, rickshaw and van pulling, scrap collection, shopkeeping and sales.

The destination of the seasonal migrants of these villages has also diversified compared to the past. Currently, they migrate to nearby villages, Satkhira district capital, Jessore, Khulna, Gopalganj, Pirojpur, Munshiganj and Dhaka. The cost of migration ranges from BDT 20 to BDT 600. The average duration of stay during each migration is 1.5 months. In the past the aggregated migration period for the international migrants in one year was five months. In recent years it has increased to as much as nine months. International migration entails huge cost.[7] In comparison, the cost of internal migration is very low. None of the villagers saw the cost of migration as a hindrance to their seasonal migration. Besides, the transportation cost of those who went to work in the brickfields was borne by labour contractors. Seasonal migration is preferable to these groups of people for both economic and social reasons. They leave behind their families who have few other income sources based in the household or village surroundings. Keeping the family in the village also allows their young members to avail themselves of basic educational opportunities and access the government's 'Food for Education Programme'. The migrant families manage their life by gaining income from both rural and urban areas and ensuring social and community benefits from the rural areas.

Since the 1990s, 5 to 10 per cent of villagers, mostly belonging to the middle and upper classes, have permanently migrated. However, after the *Sidr* and *Aila* Cyclones, 25 per cent of the villagers migrated permanently or sent their families mostly to nearby Khulna and Jessore districts. Permanent migration was an option for the relatively better off households. After *Aila* 150 families from these villages permanently migrated to India. The migrating households to India had a section of their family members already residing in different parts of West Bengal. When the environmental stress surfaced it was easier for them to decide to migrate permanently. The cost of cross-border migration from Satkhira to the West Bengal border is very low. Different types of migration are used by people of all economic strata for adapting to change. Along with the environment, social, economic and many other factors contribute to the decision to migrate.

6.6.2   Nawabganj

The study of Nawabganj covered Kheshba, Mohanohil, Durlogpur and Charpaka villages. The first two villages experience seasonal drought and the other two experience drought as well as flood and riverbank erosion. In the past, agriculture and agro-based enterprises and seasonal migration as agricultural labourers constituted the major livelihood opportunities for the villagers. These villages are experiencing extreme climatic conditions. The villagers recall that there has been a sharp drop in the groundwater level. It went down from 60–70 feet in the 1980s to 70–80 feet in the 1990s. By 2012, it had fallen to 140–160 feet. The cost of irrigation has gone up considerably and poor people cannot afford it. Production is lost frequently due to drought; 70 per cent of the village land is affected by drought. Over the years, along with agricultural work within the village, or nearby, seasonal inter-district migration for agricultural work and migration to mega-cities as construction workers and day labourers have become important livelihood options.

Interestingly, Charpaka village was part of a *Char* area that emerged over the *Padma* River in 1970. People from Ujirpur, Golakpur and Paka *Unions* of Shibganj sub-district who lost their villages to riverbank erosion resettled in Char Paka. People had grown trees and established homesteads. No work was available in the village. They found work in the neighbouring villages of India, mainly in the paddy fields, as daily commuters and temporary migrants. Some were involved in informal border trade. Since 1995, border control had become stricter and subsequently the villagers found jobs in other districts of Bangladesh as internal migrants.

The construction of Jamuna Multipurpose Bridge has facilitated migration to mega-cities such as Dhaka and Chittagong. It costs BDT 60–70 to go to Rajshahi from Nawabganj and BDT 300–500 to go to Dhaka. Local contractors called *Sardars* have connections with agricultural landlords of different districts and arrange seasonal agricultural migration. The cost of migration is borne by migrants. Construction contractors of Dhaka and other cities maintain a network of connections with informal local labour agents and construction workers' associations in Nawabganj. These networks facilitate migration for construction work. Like Satkhira, female migration from Nawabganj is limited.

Over the years, members of some families have also taken part in short-term contract migration. Such migration gained momentum during 2008–09 when Libya reopened its market after the withdrawal of sanctions on the country by the United Nations and the USA. Nawabganj has grown rapidly as a source area of international migrants in 2008–09. A process of step migration is visible here, for example, exposure to Dhaka as a construction worker, development of a formal network such as the association of construction workers, as well as the development of communication infrastructure very quickly increased the number of internal and subsequently international migrants from these villages. One can see that in facing the challenges caused by drought and riverbank erosion, people of Nawabganj are using the migration of a few family members as a way of income diversification and adaptation.

6.6.3 Munshiganj

Munshiganj is located 50 kilometres from the capital city of Dhaka. Three villages, Bhaggyakul, Mandra and Charipara of Sreenagar subdistrict, and one village, Char Sonakanda of Sirajdikhan subdistrict were studied. The three villages are on the mainland and people started living on Char Sonakanda from the 1970s.

Riverbank erosion, excessive rainfall in one season and lack of rainfall in another season as well as flooding are major environmental stresses faced in these villages. Due to riverbank erosion many villagers lost all of their land including their homesteads and some became landless and destitute. Most of the poor villagers resettled in nearby villages. Relatively better off ones settled both locally as well as in the cities. Some migrated to Dhaka and nearby districts.

Fishing and agriculture were the two major areas of employment in the past. Traditional livelihoods have changed over the years. The fishing community has shrunk significantly. Pollution, non-availability of fish, lack of capital and indebtedness have contributed to the decimation of this community. Land loss due to erosion, population density, low availability of arable land, low profit from agriculture, proximity to business centres and the scope of internal and international migration have reduced the interest of the local population in agricultural production. A section of the poor has migrated to nearby business centres as low-paid workers. Some commute to Dhaka and adjacent cities to work as day labourers, hawkers, vendors and rickshaw pullers. Employment in garment and leather manufacturing has attracted a section of working-age men and women to Dhaka, Gazipur and Narayanganj districts as permanent migrants.

Since the mid 1980s, some of the villagers have been migrating to the Middle East and Southeast Asian countries as short-term contract workers. Family members of the international migrants reside in these villages. Due to internal and international migration, there has been a labour shortage for certain types of work in the three villages of Sreenagar subdistrict. This has drawn internal migrants particularly from south-western Bangladesh and the coastal belt, such as Satkhira, Bagherhat and Jessore districts, as well as from poverty-stricken North Bengal districts of Kurigram, Gaibandha and Rangpur. The Mawa ferry terminal is only 8–10 kilometres away from these villages and the internal migrants have also found jobs in the ferry terminal. Munshiganj is a good example of the theoretical construct presented earlier.

In this scenario, it is very difficult to identify the most important contributing factor in the migration decision of individual villagers of different professions. It could be flood and loss of land due to riverbank erosion, it could be proximity to the capital city, it could be availability of information or it could be all of the above along with a household's or individual's agency that produced migration. Therefore, instead of looking at the role of climate change events in inducing migration, it is more important to learn about the role of migration in the process of adaptation.

The 12 villages of three districts experienced different types of environmental stresses and shocks. Interestingly, the people of these villages had been using migration as one of their adaptation strategies for a long time. The household income of the poorer people of the villages of Satkhira was maintained to a large extent with earnings from seasonal migration. After Cyclones *Sidr* and *Aila*, the time span of seasonal migration increased from six months to nine months. Charpaka villagers from Nawabgonj earned their livelihood by commuting to neighbouring villages of India. Once such scope was reduced, a large number of working-age men earned their livelihoods as seasonal agricultural workers or construction workers in different cities. The villagers of Munshiganj experienced both rural to urban as well as rural to international migration in search of new livelihoods in the context of loss of traditional livelihoods. Loss of some livelihood opportunities might have been caused by climate shocks and other factors.

6.7   CONCLUSIONS AND RECOMMENDATIONS

This chapter has analysed the current framework of looking at migration in the adaptation to climate change discourse of Bangladesh. The chosen conceptual framework highlights that the relationship of migration with climate change is not linear and climate change shocks alone do not contribute to migration. Social, economic, political and climate change shocks mediate with the individual's agency and presence or absence of facilitation forces produce movement of some people and non-movement of others. It is difficult to estimate the extent of the contribution of climate change to migration. However, it is possible to say that people use migration as an adaptation tool when they are faced with shocks. Nonetheless, the type of migration they undertake varies according to circumstances.

The chapter has reviewed different policies and action plans of government that deal with climate change and adaptation. It found that successive governments' initial policies and adaptation plans presented migration as a harmful impact of climate change. Government literature systematically cites the Intergovernmental Panel on Climate Change (IPCC) or figures from other institutions to present population displacement as one of the major threats of climate change. Different government policies have also linked such migration with an increase in crime in urban areas. Thus, restricting the scope of migration naturally became the major target of those policies. It is only in BCCSAP that the negative presentation of migration was jettisoned. Nonetheless, migration is perceived as a problem and the burden of responsibility has been placed on the shoulders of those who are responsible for climate change in the first place and the permanent resettlement of displaced people through international migration has been suggested.

International NGOs and civil society bodies of Bangladesh have undertaken a large number of adaptation programmes. These include agricultural innovation, generation and use of local knowledge, safe housing and infrastructural development for the reduction of risk and so on. These attempts also did not see migration as one of the strategies used by families to substitute income losses due to climatic stresses. Civil society and their development partners similarly could not depart from the perception of migration as a threat.

We have reviewed existing literature on rural–urban migration. Our review shows that a section of people in challenging situations use migration as a tool for earning income and ensuring sustenance. Many types of migration experiences have been identified. To continue to live in their habitual residence the most common form entailed the seasonal migration of a few members of the family. Which type of migration will take place from a particular area depends on many factors, such as the type of worker needed at the destination point, the availability of social networks, access to information and cost of migration.

The qualitative research conducted in 12 villages for this chapter particularly shows that people use migration as one of their adaptation strategies. A substantive section of working-age men from the relatively poorer households of four villages of Satkhira district now work outside their villages for nine months a year. Previously, they used to migrate seasonally altogether for six months. In the case of Nawabganj, villagers who resettled in Char Paka after losing their land to riverbank erosion in the 1980s found work in neighbouring Indian villages. After stricter control was imposed along the border between India and Bangladesh in 1995 and after the construction of the Jamuna Multipurpose Bridge and subsequent construction boom, members of poorer households found work in the construction sector of Dhaka and Chittagong. Such migration helped maintain the subsistence of their families in rural areas. A section of the affected households had also migrated internationally since 2008, Libya being the major destination. Riverbank erosion and floods in

Munshiganj resulted in the migration of both men and women from the area to places near Dhaka. Labour shortage due to such migration again pulled migrants to Munshigang from places like Satkhira, Rangpur and Gaibandha districts.

6.7.1 Recommendations

Experts have identified that since the 1980s it is the internal migrants working in the garment manufacturing sector and international migrants mostly working as short-term contract workers who have largely contributed to the economic growth of Bangladesh (Rahman 2010, Siddiqui and Bhuiyan 2013; Siddiqui and Billah 2012). Keeping this in mind, the perspective of looking at migration in the climate change context needs to be changed fundamentally. Instead of viewing migration as a threat, it should be seen as a form of adaptation. The knowledge gap on the use of different types of migration by those who are facing climate change shocks needs to be bridged. Robust data need to be generated on the experiences of the use of migration as one of the adaptation tools.

Case studies show that most internal migrants work in the informal sectors. A planned adaptation strategy can help these groups of migrants develop their skills to participate in formal sector jobs. Specialised skill development programmes can target the garment and other manufacturing industries, skilled jobs in the construction sectors and also hospitality and other service sectors.

A section of people from Munshiganj as well as Nawabganj are taking part in international migration. Government offices, the Bureau of Manpower Employment and Training and its District Employment and Manpower Offices as well as Technical Training Centres have minimal presence in most of the climate change-affected areas. The branches of these institutions should be extended to the affected areas.

International migration requires access to substantive financial resources. The government has established a specialized bank named *Prabashi Kalyan Bank* (Migrants' Welfare Bank). One of the important aims of this bank is to provide loans to potential migrants at a low interest for financing migration. Resources need to be allocated from the climate change trust fund to this bank so that it can serve potential migrants of climate change-affected areas.

The Ministry of Expatriates' Welfare and Overseas Employment looks after international labour migration. The Labour Ministry is in charge of looking after the rights of the workers who are employed within the country. None of these ministries are part of the inter-ministerial body that looks after the implementation of BCCSAP. Representation of these ministries has to be ensured in the management structure of BCCSAP.

The 2009 *Human Development Report* of the United Nations Development Programme (UNDP) identified migration as an important driver of human development. The United Nations Global Forum on Migration and Development (2009) and the Symposium of Global Migration Group (GMG) (May 2010) also recommended incorporating migration into national development strategies. Bangladesh should also incorporate migration in its targeted development strategies as well as one of the climate change adaptation tools in its *Five Year Development Plans* and the *Ten Year Perspective Plan*.

NAPA of the Government of Bangladesh, disaster risk reduction strategies and BCCSAP should look at migration as an option rather than as a threat. The Overseas Employment Policy of the Government of Bangladesh should incorporate the issue of climate change. Its implementation strategy should accommodate facilitation of international migration from climate change-affected areas. This will necessitate capacity building of Ministry of Expatriates' Welfare and Overseas Employment, Ministry of Labour, Bureau of Manpower Employment and Training and District Employment and Manpower Offices. A coordinated approach of all stakeholders is likely to transform challenges into opportunities for economic advancement of the climate change-affected people.

NOTES

1. HIES (2010), based on the Bangladesh Bureau of Statistics (BBS) method anchored to HIES (2005) upper poverty lines; inflation adjustment based on Housing Income and Expenditure Survey (HIES) data, not the consumer price index (CPI).
2. Of these 89 per cent, 70 per cent were from middle-income earning families.
3. 'Chars are low lying flood and erosion prone areas in or adjacent to major rivers. The Char Livelihood Programme has made a distinction between island chars, which are surrounded by water for most of the year and mainland chars which may be surrounded by water during flood seasons but for most of the year are approachable without the use of a boat' (Hodson 2006, p. 3).
4. 'Monga is a seasonal food insecurity in ecologically vulnerable and economically weak parts of north-western Bangladesh, primarily caused by an employment and income deficit before the rice grown in the monsoon season. It mainly affects those rural poor, who have an undiversified income that is directly or indirectly based on agriculture' (Zug 2006, p. 2).
5. *Bangladesh – National Environment Management Action Plan* (NEMAP) (Vol. 1 of 2, in English).
6. Unions are the smallest rural administrative and local government units in Bangladesh.
7. The average cost is US$3125 at the March 2013 exchange rate of Bangladeshi currency (Siddiqui and Bhuiyan 2013).

REFERENCES

Abrar, C.R and N. Azad (2003), *Coping with Displacement: River Bank Erosion in North West Bangladesh*, Dhaka: Refugee and Migratory Movements Research Unit (RMMRU).

Agarwal, A. (2008), 'The role of local institutions in adaptation to climate change', Paper presented at the Social Dimensions of Climate Change Meeting, Social Development Department, World Bank, Washington, DC, 5–6 March.

Ali, M.S. (2010), 'Climate change impacts on surface water: Bangladesh perspective', Paper presented at the Climate Change and Community-level Adaptation Conference, organized by the Monash Sustainability Institute, Monash University, Melbourne, 5–9 July.

Aminuzzaman, M.S. (2010), 'Environment policy of Bangladesh: a case study of an ambitious policy with implementation snag', Paper presented at the South Asia Climate Change Forum, organized by the Monash Sustainability Institute, Monash University, Melbourne, 5–9 July, available at http://www.monash.edu.au/research/sustainability-institute/asia-projects/paper_salahuddin_aminuzzaman.pdf (accessed 14 April 2013).

Angeles, G., P. Lance, J B. O'Fallon, N. Islam, A. Mahbub and N.I. Nazem (2009), 'The 2005 Census and mapping of slums in Bangladesh: design, select results and application', *International Journal of Health Geographics*, **8**, 32–8.

Arongo, J. (2004), 'Theories of international migration', in D. Joly (ed.), *International Migration in the New Millennium: Global Movement and Settlement*, Research in Migration and Ethnic Relations Series, Farnham, UK: Ashgate Publishing, pp. 15–34.

BCAS (2009), *Community-based Adaptation to Climate Change* 3rd International Conference, 18–24 February, Dhaka

Chowdhury, S., A.M. Mobarak and G. Bryan (2009), 'Migrating away from a seasonal famine: a randomized intervention in Bangladesh', Human Development Research Paper (HDRP) Series 41, available at http://hdr.undp.org/en/reports/global/hdr2009/papers/HDRP_2009_41.pdf (accessed 20 October 2012).

Dolan, A.H. and J. Walker (2006), 'Understanding vulnerability of coastal communities to climate change related risks', *Journal of Coastal Research*, Special Issue, **39** (Proceedings of the 8th International Coastal Symposium (2004), Vol. III, Itajai, SC-Brazil), 1316–23.

Foresight (2011), *Migration and Global Environmental Change*, Final Project Report, London: UK Government Office for Science.

GoB (1995), *National Environment Management Action Plan*, Dhaka: Ministry of Environment and Forests, available at http://www.thegef.org/gef/sites/thegef.org/files/documents/multi_page.pdf (accessed 18 September 2012).

GoB (2007), *Consolidated Damage and Loss Assessment, Lessons Learnt from the Flood 2007 and Future Action Plan (Executive Summary)*, Dhaka: Disaster Management Bureau, Ministry of Food and Disaster Management and Comprehensive Disaster Management Programme, available at http://www.dmb.gov.bd/reports/Executive%20Summary-Flood%20Report.pdf (accessed 12 June 2012).

GoB (2009), *Bangladesh Climate Change Strategy and Action Plan 2009*, Dhaka: Ministry of Environment and Forestry, available at http://www.moef.gov.bd/climate_change_strategy2009.pdf (accessed 18 September 2012).

GoB (2013), *Bangladesh Demographic and Health Survey 2011*, Dhaka: National Institute of Population Research and Training (NIPORT) and Mitra & Associates, available at http://www.measuredhs.com/pubs/pdf/FR265/FR265.pdf (accessed 2 February 2013).

Grothmann, T. and A. Patt (2005), 'Adaptive capacity and human cognition: the process of individual adaptation to climate change', *Global Environmental Change*, **15** (3), 199–213.

Haque, C.E. and M.Q. Zaman (1989), 'Coping with river bank erosion and displacement in Bangladesh, survival strategies and adjustments', *Disasters*, **13** (4), 300–14.

HIES (2005), *Household Income and Expenditure Survey*, Dhaka: Statistics Division, Ministry of Planning, Government of Bangladesh.

HIES (2010), *Preliminary Report on Household Income & Expenditure Survey*, Dhaka: Statistics Division, Ministry of Planning, Government of Bangladesh.

Hodson, R. (2006), *The Char Livelihood Programme: The Story and Strategy So Far*, Dhaka: DFID, available at http://www.bracresearch.org/publications/clp.pdf (accessed 12 August 2012).

Hugo, G. (1996), 'Environmental concerns and international migration', *International Migration Review*, **30** (1), 105–31.

Hutton. D. and C.E. Haque (2004), 'Human vulnerability, dislocation and resettlement: adaptation processes of river-bank erosion-induced displaces in Bangladesh', *Disasters*, **28** (1), 41–62.

ILO (2008), *Cyclone Sidr: Preliminary Assessment of the Impact on Decent Employment and Proposed Recovery Strategy*, Dhaka: ILO in collaboration with the Ministry of Labour and Employment, Government of Bangladesh.

IOM (2008), 'Discussion note: migration and the environment', MC/INF/288, 1 November, Ninety Fourth Session, International Organization for Migration, Geneva, 14 February 2008, available at http://www.iom.int/jahia/webdav/shared/shared/mainsite/about_iom/en/council/94/MC_INF_288.pdf (accessed 10 April 2013).

Kniveton, D., K. Schmidt-Verkerk, C. Smith and R. Black (2008), *Climate Change and Migration: Improving Methodologies to Estimate Flows*, Geneva: International Organization for Migration.

Kniveton, D. R., C.D. Smith, R. Black and K. Schmidt-Verkerk (2009), *Challenges and Approaches to Measuring the Migration – Environment Nexus*, Geneva: International Organization for Migration.

Martin, M., M. Billah, T. Siddiqui, D. Kniveton and R. Black (forthcoming a), 'Climate change, migration in rural Bangladesh: a behavioural model', *Journal of Population and Environment*.

Martin, M., Y.H. Kang, T. Siddiqui, C. Abrar, K. Kniveton and R. Black (forthcoming b), 'Climate-influenced migration in Bangladesh: the need for a policy realignment', *Journal of Development Policy Review*.

Massey, D.S., J. Arango, G. Hugo, A. Kouaouci, A. Pellegrino and J.E. Taylor (1998), *Worlds in Motion, Understanding International Migration at the End of the Millenium*, Oxford: Clarendon Press.

McCarthy, J., O. Canziani, N. Leary, D. Dokken and K. White (eds) (2001), *Climate Change 2001: Impacts, Adaptation and Vulnerability*, Contribution of Working Group II to the Third Assessment Report of the Intergovernmental Panel on Climate Change, Cambridge: Cambridge University Press.

Paul, B.K. (2003), 'Relief assistance to 1998 flood victims: a comparison of the performance of the government and NGOs', *Geographical Journal*, **169**, 75–89.

Rahman, H.Z. (2010), *Bangladesh: Strategy for Accelerating Inclusive Growth*, Dhaka: Power and Participation Research Centre (PPRC).

Rahman, A.A., M. Alam, S.A. Uddin, M. Rashid and G. Rabbani (2007), 'Risks, vulnerability and adaptation in Bangladesh', Background paper of UNDP *Human Development Report 2007*, available at http://hdr.undp.org/en/reports/ global/hdr2007-8/papers/Rahman_Alam_Alam_Uzzaman_Rashid_Rabbani. pdf (accessed 14 July 2012).

Rayhan, I. and U. Grote (2007), 'Coping with floods: does rural-urban migration play any role for survival in rural Bangladesh', *Journal of Identity and Migration Studies*, 1 (2), 82–98.

Siddiqui, T. (2009), 'Climate change and population movement: the Bangladesh case', Paper presented at the Climate Insecurities, Human Security and Social Resilience Conference, RSIS Centre for Non-traditional Security Studies, Singapore, 27–28 August.

Siddiqui, T. (2010a), 'Climate change and community adaptation in Bangladesh', Paper presented at the Climate Change and Community-level Adaptation Conference, organized by the Monash Sustainability Institute, Monash University, Melbourne, 5–9 July.

Siddiqui, T. (2010b), 'Climate change and human security', Paper presented at the Fourth Annual Convention of the Consortium of Non-traditional Security, Singapore, 25–26 November.

Siddiqui, T. (2011), 'Policies and interventions relevant to environmental migration: migration as an adaptation strategy', Paper presented at the International Workshop on Bangladesh, Low Elevation Coastal Zones and Island, organized by Foresight, UK Government office for Science and RMMRU, University of Dhaka, Dhaka, 3–4 February.

Siddiqui, T. and R. Bhuiyan (2013), 'Emergency return of Bangladeshi migrants from Libya', NTS Working Paper No.7, RSIS Centre for Non-traditional Security (NTS) Studies, Singapore.

Siddiqui, T. and M. Billah (2012), *Labour Migration from Bangladesh 2011: Achievements and Challenges*, Dhaka: Refugee and Migratory Movements Research Unit.

Skinner, J. and T. Siddiqui (2005), 'Labour migration from chars: risks, costs and benefits', Mimeo, Refugee and Migratory Movements Research Unit, Dhaka.

Sterrett, C. (2011), *Review of Climate Change: Adaptation Practices in South Asia*, Melbourne, OXFAM, available at http://www.oxfam.org/sites/www.oxfam. org/files/rr-climate-change-adaptation-south-asia-161111-en.pdf (accessed 16 November 2012).

UN (2010), *Cyclone AILA: Joint UN Multi-sector Assessment and Response Framework*, available at http://www.lcgbangladesh.org/derweb/Needs%20 Assessment/Reports/Aila_UN_AssessmentFramework_FINAL.pdf (accessed 16 September 2012).

UNDP (2009), *Human Development Report 2009, Overcoming Barriers: Human Mobility and Development*, New York: UNDP.

UNDP (2012), 'Bangladesh: country profile', *International Human Development Indicators*, available at http://hdr.undp.org/xmlsearch/reportSearch?y=*&c=n %3ABangladesh&t=*&lang=en&k=&orderby=year (accessed 12 December 2012).

UNFCCC (2011), *Report of the Conference of the Parties on its Sixteenth Session*, held in Cancun from 29 November to 10 December 2010, Addendum, Part Two: Action taken by the Conference of the Parties at its Sixteenth Session.

Walsham, M. (2010), *Assessing the Evidence: Environment, Climate Change and Migration in Bangladesh*, Dhaka: International Organization for Migration.

Zug, S. (2006), 'Monga – seasonal food insecurity in Bangladesh – bringing the information together', *Journal of Social Studies*, **111**, available at http://www.bangladesch.org/pics/download/S_Zug_Article_Monga.pdf (accessed 15 September 2012).

7. Adaptation strategy to address climate change impacts in the mountains: the case of Nepal

Bhaskar Singh Karky*

7.1 INTRODUCTION

The Mountains of the Hindu Kush Himalaya (HKH) region are known as water towers or the Third Pole as they bear more ice and snow than any other region outside the Poles (Dyhrenfurth 1955; Qiu 2008). They are one of the most vulnerable regions from a climate change perspective. Mountains are considered reliable indicators of global warming and provide an opportunity to better understand climate change impacts despite inadequate information.

This chapter focuses on the mountain systems of the HKH region to better understand climate change impacts and vulnerability. It takes the case of Nepal, a mountainous country of the Himalayas, to discuss adaptation needs, and draws on key learning for a mountain-specific adaptation strategy. The chapter analyses government policies and programmes and climate initiatives to illustrate policy responses and effectiveness of adaptation to climate change.

7.1.1 Mountain Ecosystem Services

Globally, mountains cover 24 per cent of the Earth's terrestrial area (UNEP WCMC 2002) and are home to approximately 12 per cent of the population (Huddleston et al. 2003). Mountains are a significant repository of ecosystem services – they are originating sources for all the major rivers and the supply of fresh water to downstream areas (Viviroli et al. 2007). They are rich in biodiversity as well as cultural diversity. Mountains are also providers of hydroelectricity and are greatly valued as a tourism and pilgrimage destination.

The HKH region is a source of ten major river systems – Amudarya, Tarim, Indus, Ganga, Brahmaputra, Irrawady, Salween, Mekong, Yangtse and Yellow. These rivers provide fresh water to their densely populated basins across Asia and are also an important source of hydropower for sustaining Asian economic growth. This range spans over 3500 km of terrain with long stark changes in elevation that result in an extremely diverse biogeographic region ranging from coastal mangroves to arid mountain deserts. For instance, this region has '488 protected areas (IUCN [International Union for Conservation of Nature] Category I-VI); 29 Ramsar Sites; 13 UNESCO [United Nations Educational, Scientific and Cultural Organization] Heritage Sites; 330 Important Bird Areas' (Singh et al. 2011, p. 24). This region is not only extremely vulnerable to risks from climate change but also has over 40 per cent of the world's poorest population (ICIMOD 2012).

7.1.2   Mountain Development and Vulnerability

Although mountains provide necessary ecosystem services for people, the mountain environment is highly vulnerable to different perturbations. Planners consider geography as the most important aspect in development. Literature on comparative growth between mountain and non-mountain countries has indicated that economic development is very much influenced and shaped by geophysical and agro-ecological factors (Gallup et al. 1999; Masters and McMillan 2001). They show that countries with mountains are characterized by differences in terms of resource endowment, mobility of production factors and agricultural productivity, and consequently their infrastructure and human capital development require more resources. The sheer remoteness, isolation and limited infrastructure become obstacles for realizing food security and economic development in mountains (Akramov et al. 2010; Huddleston et al. 2003).

These features of mountains that hinder economic and overall development are referred to as 'mountain specificities' by Jodha (1992, pp. 44–45). The specificities that hinder development and economic growth are classified as inaccessibility, marginality and fragility. The HKH region provides good evidence of how mountain specificities stand out in causing the region to lag behind in economic development. In this context, mountain specificities become a subject of rigorous analysis for overcoming the constraints for socio-economic development and proliferating opportunities associated with them.

Not only are the mountain regions lagging behind in economic development due to mountain specificities, but are also being increasingly recognized as highly vulnerable to the impact of climate change. The HKH region is regarded as the region most vulnerable to climate change globally as ranked by Maplecroft (2012) in the new Climate Change Vulnerability

Index (CCVI). Out of the 170 countries in the CCVI ranked over the next 30 years, several that are affected by the HKH mountains are ranked very high: Bangladesh (1), India (2), Nepal (4), Afghanistan (8), Myanmar (10) and Pakistan (16). It is estimated that more than 40 per cent of the world's poorest people live in the river basins that originate from the HKH region (ICIMOD 2012). As Nepal represents in miniature the greater Himalaya region, a careful study of Nepal's mountain contexts not only helps understand the country's vulnerability but also generates useful learning for the rest of the region that depends on ecosystem services generated from the mountains.

7.2 PHYSIOGRAPHY OF NEPAL

Nearly 83 per cent of Nepal's topography is mountainous and gives rise to unique and diverse physiographic and climatic conditions as shown in Figure 7.1. The total country area is around 147,000 km^2 with five distinct physiographic regions based on elevation (LRMP 1986): high Himal, high mountain, middle mountain, Siwalik (or the Churia) and the Terai (or the plains) as shown in Table 7.1. Table 7.1 also shows that half the population

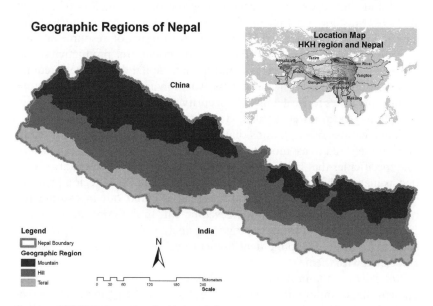

Source: ICIMOD. Reproduced with permission.

Figure 7.1   Map of Nepal showing different geographic regions

Table 7.1   Physiographic zone showing population distribution and poverty incidence in Nepal

Physiographic zone	Ecological belt	Climate	Average annual precipitation (mm)	Mean annual temperature (°C)	Population (year 2011) (million)	Poverty incidence (year 2005)
High Himal High mountain	Mountain	Arctic/ alpine	Snow/150–200	< 3–10	1.8	0.425
Middle mountain	Hill	Cool/ warm	275–2300	10–20	11.4	0.336
Siwalik Terai	Plains/ Terai	Tropical/ sub-tropical	1100–3000	20–25	13.3	0.295

Source: WECS (2005) and ISRC (2013).

resides in the mountain and hill region and the remainder in the Terai plains. The incidence of poverty is higher in the hills and mountains than the plains, indicating that widespread poverty is concentrated in the uplands.

7.2.1   Scientific Basis of Climate Change

Observed changes
Several studies have confirmed that there is consistent warming and rise in the maximum temperature occurring more at higher altitudes. Studies by Shrestha et al. (1999) and Shrestha and Aryal (2011) collected temperature data from 1977 to 2011 from 49 stations spread over Nepal that is published in Eriksson et al. (2009, p. 5) (Table 7.2).
The results indicate that:

- The average maximum temperature between 1977 to 2000 increased at a rate of 0.06°C per year for Nepal.
- The rise in temperature was greater at higher altitudes, trans-Himalayan region, middle mountains and Himalaya.
- The rise in temperature between 1977 to 2000 occurred mainly in the cooler months (0.06–0.08°C per year from October–February, for all of Nepal).

These results are consistent with other studies of Nepal. Kadota et al. (1997) estimated a 1.4°C temperature rise from 1989 to 1991 in Shorong Himal at 4958 m above sea level, which is close to the observations of Shrestha et al. (1999).

Table 7.2　Observed trend for temperature in Nepal

Region	Mean Max Temperature Trends for the period 1977–2000 (°C per year)				
	Seasonal				Annual
	Winter December– February	Pre- monsoon Mar–May	Monsoon June– September	Post-monsoon October– November	January– December
Trans- Himalaya	0.12	0.01	0.11	0.10	0.09
Himalaya	0.09	0.05	0.06	0.08	0.06
Middle mountains	0.06	0.05	0.06	0.09	0.08
Siwalik	0.02	0.01	0.02	0.08	0.04
Terai	0.01	0.00	0.01	0.07	0.04
All Nepal	0.06	0.03	0.05	0.08	0.06

Source:　Eriksson et al. (2009), p. 5.

In a separate study, Baidya et al. (2008) analysed the daily temperature data for 36 years from 1971 to 2006 and also revealed that most of the temperature extreme indices show a consistently different pattern in the mountains and the Terai. They show that there is a decreasing trend of cool days and an increasing trend of warm days that are more evident at higher altitudes and this trend has a higher magnitude in the mountains. This is also consistent with McSweeney et al. (2012), which states that the projected rate of warming is most rapid in December, January and February (DJF) and March, April and May (MAM).

There is no clear trend for changes of precipitation in Nepal (Shrestha et al. 2000). The *National Adaptation Programme of Action* (NAPA) (2010) (MOEnv 2010) mentions that the annual precipitation data reveals a general decrease in pre-monsoon rainfall in western Nepal and an increase in the east. Monsoon precipitation also shows a decline in the west but an increase in the rest of the country.

A study undertaken by Baidya et al. (2008) highlights that precipitation extremes (heavy showers) are increasing though there is no systematic difference between the Terai and mountain regions. The study notes that there is an indication of increased frequency of weather-related extreme events such as floods and landslides.

Projections of future climate

There have been few climate projection studies for Nepal, mainly due to lack of resources, capacity and technology for climate modelling. The Organisation for Economic Co-operation and Development (OECD) carried out an assessment of 12 recent General Circulation Models, of which seven were run based on the Special Report on Emission Scenarios (SRES) B2 (OECD 2003). SRES B2 refers to one out of the six families of scenarios discussed by the Intergovernmental Panel on Climate Change (IPCC), characterized by population growth, medium economic development and less rapid technological change with a regional focus, and based on this SRES scenario future greenhouse gas concentrations are projected. Using the B2 scenario, the results indicate significant and consistent warming in winter months. The projected change above the baseline average is 1.2°C for 2030, 1.7°C for 2050 and 3.0°C for 2100. For precipitation, there is less significant change and high standard deviation among the model results.

A downscaled Global Climate Model (GCM) for Nepal is being analysed by the United Nations Development Programme (UNDP) as part of the initiative to develop climate change profiles for developing countries. The study relies on multi-model projections from the World Climate Research Programme (WCRP) Coupled Model Intercomparison Project (CMIP3) archive and updated existing climate observations to generate country-level data plots. The major findings for temperature and precipitation projection are shown in Box 7.1.

7.2.2   Climate Change Impacts in Nepal

Agriculture and food security

The agriculture sector of Nepal has not modernized but it provides a broad base for the country's economy. As indicated in Figure 7.2, the agriculture sector alone contributes 38 per cent of the gross domestic product (GDP) and around 78 per cent of households in the country work in this sector (NLSS 2004).

Of the agricultural holdings, 90 per cent are in the hills and Terai and 10 per cent in the mountains. Of the total agricultural land, only 54 per cent is irrigated, with the remaining subject to the vagaries of nature (NLSS 2004). There is a dearth of finance and technological investment in the agriculture sector with an average landholding size of 0.83 hectare, decreasing from 1.09 hectares in 1995 (NLSS 2004). Food and Agriculture Organization (FAO) data shows that in 2008, 17 per cent of the population was undernourished (FAO 2012).

According to a World Food Programme Nepal (WFPN 2009) report,

---

BOX 7.1   NEPAL: UNDP GCM PROJECTION OF FUTURE CLIMATE CHANGE

Temperature

- The mean annual temperature is projected to increase by 1.3–3.8°C by the 2060s, and 1.8–5.8°C by the 2090s. The range of projections by the 2090s under any one emissions scenario is 1.5–2°C.
- The projected rate of warming is most rapid in DJF and MAM.

Precipitation

- Projections of mean annual rainfall averaged over the country from different models in the ensemble are broadly consistent in indicating increases in rainfall over Nepal. This is largely due to increases in JJA and SON (wet season) rainfall.
  - JJA rainfall is projected to change by –36 mm (–22 per cent) to +224 mm (+104 per cent) per month by the 2090s. SON rainfall is projected to change by –17 mm (–38 per cent) to +44 mm (+71 per cent) per month by the 2090s. These increases are offset a little by projected decreases in DJF rainfall, such that annually, projected changes range from –14 mm (–31 per cent) to +59 mm (58 per cent) per month.
  - The increase in JJA rainfall is largest in the south-east of Nepal.

Note: JJA – June, July, August; SON – September, October, November; DJF – December, January, February.

Source: McSweeney et al. (2012), pp.2–3. Reproduced with permission.

Nepal has a severe food deficit and faces widespread hunger. The situation is alarming as the conditions in the mountains and hills of western Nepal are comparable to Ethiopia (WFPN 2009). The higher elevations in the western region are the most food-deficit areas and rely on rice subsidies from the Nepal Food Corporation as there is little scope for improving the fertility of arable land. A large proportion of the population depends on agriculture, with very small landholdings, half of which lack irrigation and

Source: World Bank (2012).

Figure 7.2 Contribution of the agriculture sector to Nepal's GDP

mainly provide a subsistence livelihood (Figure 7.3). This exposes the scope of the vulnerability of Nepal's agriculture and food security to many additional drivers of change such as migration, the feminization of mountain agriculture, land degradation, weak local governance, lack of investment in the agriculture sector in addition to climate change. Though climate change is not a major driver, it further intensifies the existing vulnerability.

The impact of climate change in the agriculture sector is mostly evident as a result of extreme weather events. There needs to be more research in this area as the impacts of different climate scenarios have not been analysed sufficiently with local-level data. But water stress is one of the most significant adverse impacts of climate change as indicated by many farmer perspective studies (ongoing International Centre for Integrated Mountain Development-Institute for Global Environmental Strategies (ICIMOD-IGES) study). Water stress leads to reduced Net Primary Productivity (NPP) and has a direct bearing on food production, especially for subsistence farmers who have few adaptation options. NPP is the scientific measure of the rate at which plants convert water, sunlight and carbon dioxide into glucose and oxygen.

Note:　The mountain farming system is characterized by traditional practices in small parcels of land that are rain fed. In the background is the receding snow line of the Gangapurna glacier in Manang district.

Source:　Author.

Figure 7.3　Nepal's mountain farming system

Table 7.3 shows the change in population and Food Production Index of five countries in South Asia. Nepal is the only country to have a decreasing rate of change in annual percentage in Food Production Indices, decreasing from 0.66 between 1990–92 to 2000–02 to 0.34 in 2000–02 to 2006–08. Nepal and Pakistan are the only countries where the growth rate in the Food Production Index has not been able to catch up with the population growth rate.

Nepal is a food importer. Its population depends on many staples procured from India. Food pricing, subsidy policy, agriculture technology and drought in India determine the affordability and availability of food in the food-deficit hills and mountains of Nepal. From a climate change perspective, a 1°C rise in temperature in the Indian subcontinent will decrease 4–5 million tonnes of wheat production (GIZ 2011, p. 18), which will also increase the food insecurity of the poor populations of Nepal.

Table 7.3   Percentage change in population and Food Production Index in South Asian Countries

Country	Annual change in population		Annual change in percentage in Food Production Index	
	1990–92 to 2000–02	2000–02 to 2006–08	1990–92 to 2000–02	2000–02 to 2006–08
Bangladesh	2.13	1.68	1.31	2.63
India	2.05	1.64	0.54	1.70
Myanmar	1.33	0.76	5.00	9.11
Nepal	2.77	2.19	0.66	0.34
Pakistan	2.77	2.37	0.91	1.39

Source: WHO (2012).

Water resources and hydropower

Nepal typically gets nearly 80 per cent of its annual rainfall between June and September, and 20 per cent in the remaining eight months. Climate change directly affects the spatial and temporal distribution of fresh water. It can especially intensify drought in the drier season between March and May. These impacts are exhibited in mountain regions as well as the low lands downstream. The steep mountain slopes that produce orographic precipitation and accumulate substantial amounts of snow and ice in several hundred glaciers and mountain tops are highly sensitive to the temperature changes. They are inherently vulnerable to a rise in temperature and this poses challenges to water resources management. Seasonal melting of snow from the Himalaya region plays a crucial role in balancing the water deficit in the long, dry summer. Climate change, however, is very likely to worsen this delicate balance by widening the gap between water demand and supply, affecting the large population living in the Ganga Brahmaputra Meghna (GBM) basin where water stresses are already critical.

Nepal's climate, predominantly influenced by the monsoons and westerly disturbances, comprises summer, rainy season and winter. Westerly disturbances are storm systems that originate in the Mediterranean and are responsible for rain and snow during winter, which is much needed for the winter crops, also called rabi crops. Seasonal and spatial variability of the annual snow and ice melt, including precipitation contribution to discharge, are uncertain in this region (Bolch et al. 2012). Climate change impacts are already evident in floods, drought, debris flows, vector- and water-borne diseases, forest fires and degradation of ecosystems. Owing to its geophysical characteristics, Nepal is particularly susceptible to the impacts of climate change on water resources due to its low water storage capacity and low productive water use. Nepal's climate projections indicate that precipitation will be further skewed in the coming decades. In the context of a stable climate, snowmelt contributions could have helped ameliorate the imbalance of water supply during the drier months, but is no longer possible.

The hydropower sector in the HKH region is a priority sector of the governments as it is a major driver of economic growth and also a significant source of revenue in Nepal, Bhutan and northeast India. Water resources and hydropower are directly affected by rising temperatures, which causes faster glacier retreat, leading to a diminishing flow of streams in the long, dry summer, thus having a negative impact on hydropower generation. Climate change is causing greater variability in precipitation and more frequent droughts, leading to reduced stream flow. Other climate-induced disasters and risks to hydropower facilities include glacial lake outburst flooding (GLOF), flooding, landslides, sedimentation and increased and unreliable dry season flows. In the Brahmaputra and Kosi basins, Gosain et al. (2011) have predicted 25–40 per cent more sedimentation even if there is no land use change in the river basin, indicating the vulnerability of the hydropower sector to climate change impacts.

The ten-year *Hydropower Development Plan* (MOE 2009) has projected the requirement of 2112 MW, 2882 MW and 4990 MW of electricity by 2020 under 'business as usual', 'medium' and 'high' growth scenarios, respectively. The *Energy Emergency Plan* (2011) proposed the need to generate 2500 MW of electricity by 2015 (MOE 2009). The impact of climate change on water availability and water-induced disasters will be key challenges to this plan.

Much of the hydropower and irrigation facilities in the country are run-of-river types that yield low productive water use and are susceptible to water-related disasters. Nepal does not have infrastructure developed to support water storage except for one reservoir, Kulekhani, which stores the monsoon rain to be utilized during the dry season (Figure 7.4). To augment the uncertain meltwater regime that affects hydropower generation and irrigation, more reservoir projects need to be developed in the country especially if the *Hydropower Development Plan* is to be realized.

Biodiversity
Nepal's altitude ranges from 60 m to the highest vegetation line with climatic variations that enable the country to harbour diverse critical ecosystems (NBS 2002). Nepal 'possesses a disproportionately rich diversity of flora and fauna at ecosystem, species, and genetic levels' on less than 0.1

Source: Author.

Figure 7.4   Kulekhani reservoir: a 92 MW water storage hydro plant

per cent of total land area of the world (MOEST 2008, p. 4). A total of 118 ecosystems have been identified in different physiographic zones of Nepal (REDD-Forestry and Climate Change Cell 2011).

Maskey (1996) in REDD-Forestry and Climate Change Cell (2011, p. 3) reports that '26 species of mammals, nine birds and three reptiles are either endangered or vulnerable or threatened and include tiger, rhinoceros, elephant, musk deer, snow leopard, swamp deer, wild buffalo, bengal florican, lesser florican, red panda, clouded leopard, gangatic dolphin, gharial'. These species are threatened by loss of their habitat mainly due to human activity and their vulnerability is further intensified by climate change. The prolonged dryness in the hills of Nepal in 2009 and 2010 saw a rise in forest fire incidence throughout the country. Such impacts from climate change adversely affect biodiversity directly.

For example, snow leopard is a keystone species of the high altitude Himalayas. It is a species requiring a large territory in sub zero temperatures at high elevations. Climate observations have shown warming in high elevations is occurring at a faster rate than in lower elevations and the projections for the future also show the same trend (Shrestha and Aryal 2011;

Shrestha et al. 1999). This means that this species may get squeezed out as there is no higher ground to move to. Forrest et al. (2012, p.129) state that 'about 30 per cent of snow leopard habitat in the Himalaya may be lost due to a shifting treeline and consequent shrinking of the alpine zone, mostly along the southern edge of the range and in river valleys', which includes Nepal's Himalaya region.

7.3 ADAPTATION STRATEGIES

Climate change-related interventions and activities are relatively new in development planning in Nepal. The Ministry of Science, Technology and Environment (MOSTE), formally known as the Ministry of Environment, Science and Technology (MOEST), has been active in developing policies and programmes in response to climate change as the adverse impacts are already evident in many areas.

There is no national legislation that explicitly addresses the regulatory responses to climate change. However, there are several climate change-related national policy documents:

- Climate Change Policy (2011)
- NAPA (2010)
- Local Adaptation Programme of Action (LAPA) Framework
- Mainstreaming Climate Risk Management in Development (Strategic Climate Fund – Pilot Programme for Climate Resilience, PPCR)
- Working on National Reducing Emission from Deforestation and Forest Degradation (REDD+) Strategy.

The Climate Change Policy (2011) was specifically formulated as nego-tiations under the United Nations Framework Convention on Climate Change (UNFCCC) progressed and a need was felt for a policy instru-ment to develop a national agenda in line with the global climate agenda and reflect the way forward for the country in dealing with climate change, particularly in paving the way for implementing mitigation and adaptation projects. The policy recommends adopting a low-carbon national develop-ment pathway by pursuing a climate-resilient approach for socio-economic development. This policy reflects the country's vision in dealing with and adapting to the adverse impacts brought about by climate change across different sectors and suggests the establishment of a Climate Change Centre.

NAPA was formulated mainly for the purpose of making Nepal eligible for adaptation finance under the UNFCCC as it is a requirement for least developed countries (LDCs). It identifies and prioritizes urgent and immediate actions for adaptation to climate change and has identified six thematic sectors and nine programmes requiring an estimated US$350 million. (Refer to http://www.napanepal.gov.np/ to see the nine programmes and their specific activities.) It is unclear on the estimation method, but NAPA commits 80 per cent of this fund to be expended at the field level (village/municipal) through a designated line ministry for implementation, which is very sound in principle.

NAPA is implemented through the LAPA Framework, which is a framework that aims to dovetail the development process with adaptation activity at the local level. Currently, the Asian Development Bank (ADB) supports the PPCR using the Climate Investment Fund and the UK Department for International Development (DFID) supports Climate Adaptation Design and Piloting Nepal (CADPN), which is piloting and testing the implementation of the LAPA Framework in selected districts in different sectors including agriculture, forest, water and sanitation, among others, by integrating adaptation planning into the District Development Committees' (DDCs) development planning process. The Programme for Scaling Up Renewable Energy in Nepal (SREP-Nepal) and the International Union for Conservation of Nature (IUCN) and UNDP ecosystem-based adaptation projects are other integrated adaptation and local development projects being launched in the country. For all adaptation-related activities, project planning and delivery is the responsibility of the Coordination Committee of the DDC.

The District Development Committee (DDC) is the apex of the local government in Nepal's administration structure. The Village Development Committees (VDCs) and, in urban areas, the municipalities are the next level of administration. There are 75 DDCs in the country that are further divided into VDCs. DDCs are responsible for development and planning at the district level, including at the village level.

In order to mainstream climate change adaptation into the development process, the Climate Resilient Planning Tool of the National Planning Commission (2010) is being used in the current three-year plan, where adaptation to climate change is mainstreamed to some extent. Similarly, the REDD Cell under the Ministry of Forest and Soil Conservation (MOFSC) is working with the Forest Carbon Partnership Facility (FCPF) of the World Bank to develop the National REDD Strategy. The REDD Cell is also developing the REDD package for the World Bank so that it can access funds to pilot REDD in the country. There are few other programmes in the forest sector such as the Norad-supported pilot REDD project ending in mid 2013 and Hariyo Ban supported by the United States

Agency for International Development (USAID), both of which are not explicitly under the government's climate change portfolio but address a range of climate concern issues.

7.3.1   Policy Gaps in Integrating Development and Adaptation to Climate Change in Nepal

Donor driven

A report by the Climate Public Expenditure and Institutional Review (CPEIR) (2011, pp. 53–4) describes a major issue with climate finance: 'A significant sum of Technical Assistance, in the order of about $13 million per year, in respect of climate related expenditure is not budgeted or accounted for through government systems (i.e., 'off-budget'). This contributes to a fragmentation of budget implementation and hinders full co-ordination of expenditure to facilitate the best effect in terms of outputs and outcomes.' This is mainly because there is a strong influence of donor finance on projects. Climate finance is frequently double counted as both an Official Development Assistance (ODA) and a climate finance pledge.

Definition of climate finance

For planning, what qualifies as adaptation expenditure is not clear due to a lack of a clear working definition for adaptation finance. A report for Nepal by CPEIR (2011) states that 'within the Government Classification Chart there is little explicit recognition of climate change-related expenditure'. However, when doing rough estimates, the same report estimates that all climate change-related activities constitute approximately 2 per cent of Nepal's GDP and around 6 per cent of government expenditure.

Differing priorities between development and adaptation

At the local level, plans are developed at the DDC level under the Ministry of Local Development (MOLD). MOLD is the coordinating agency of 75 DDCs and nearly 3800 VDCs and 60 plus municipalities. Local-level adaptation planning as per NAPA and LAPA needs to be mainstreamed in DDC budget planning. However, within the DDC budget planning, serious challenges lie in understanding and integrating climate change adaptive development activities with regular development planning. This often leads to ignoring adaptation-focused and climate-resilient development planning in favour of business-as-usual planning. In addition, climate change is still viewed as an environmental issue and not as one directly linked with the economy and development.

Inadequate capacity and institutions at local levels
At many of the remote DDCs there is an acute shortage of skilled man-power for planning development activities. This is further aggravated by lack of institutional mechanisms dedicated to integrate climate change adaptation in development. Efforts are underway to establish focal points for climate change and energy in the DDCs to strengthen mainstreaming.

Lack of institutional coordination
There is a clear lack of coordination channels between the MOSTE and DDC. MOSTE being the focal point for climate change, all finance related to climate change is channelled through this ministry to other DCCs and sectoral ministries. One reason for this is that MOSTE is not supported by district-level offices and therefore its presence is not felt at local levels. Furthermore, the DDCs' overall capacity has been undermined due to lack of elected representatives since for over a decade the VDC and DDC offices have been run on an ad hoc basis by political consensus of the few parties. This adversely affects any local-level project planning and implementation.

Difference between articulation and implementation
The Climate Change Policy aims to deliver and spend 80 per cent of the climate adaptation finance at local level. But there are lack of programmes and projects that conform to this and there is lack of capacity of public financial management to monitor this. An example is that the very donor agency that provided financial assistance to the Government of Nepal to draft the policy cannot, or has never been able to, comply with this policy.

Low carbon development pathway is referred to as the preferred devel-opment pathway in the Climate Change Policy but has not yet been main-streamed into the DDC planning process. Adaptation planning is not seen in the wider context of economic development planning.

7.4   ADAPTATION FINANCE

NAPA prioritizes several adaptation programmes for reducing vulnerabil-ity to climate change (MOEnv 2010). It proposed a US$350 million budget for urgent climate-driven adaptation requirements but has so far been successful in securing only US$10 million through the Least Developed Countries Fund (LDCF) and an additional US$25 million from bilateral aid agencies. This reiterates the dependency on donor funds for NAPA implementation.

NAPA focuses on awareness raising, capacity building and technol-ogy transfer. Several technologies are already available that are useful for climate change adaptation and vulnerability reduction. But information on the level of adoption of these technologies is scanty and the dissemination of these technologies is slow.

Adaptation strategies are crucial for managing livelihoods and natural resources in mountain areas. Nepal as a LDC with mountain features faces additional constraints for development and does not have the basic capacity to generate knowledge and information required to address immediate and pressing challenges posed by climate change, which makes it difficult to develop long-term plans for adaptation. At the global level, as a response to climate change, there are many financial mechanisms aimed at assisting the LDCs to adapt to climate change (Schwank et al. 2010). For example, the Copenhagen Accord pledged US$30 billion for adaptation and mitigation activities in developing countries from 2010–12. In addition, US$100 billion a year by 2020 has been promised through adaptation financing instruments such as the LDCF and Green Climate Fund (GCF), among others.

Several constraints are identified for financing adaptation programmes. Funding agencies often emphasize additionality in business-as-usual development programnes for financing development plans. Thus, agencies often leave the plans, citing they lack 'additionality', which is difficult to justify for specific climate-related activities because they are difficult to distinguish from business-as-usual development activities (GoN and ICIMOD 2010, p. 6). The challenge for health-related programmes has been summarized as follows: 'Climate change will make the impacts of water-borne and vector borne diseases worse, but as health sector programmes already exist they are not considered additional, despite the fact that they are clearly in need of financial assistance' (GoN and ICIMOD 2010, p. 6). Such gaps in adaptation finance will worsen poverty and slow the rate for improvement in human wellbeing. Long-term development goals are not addressed by adaptation finance as they are not additional in climate change terms.

There is concern about unintended consequences of requiring additionality as governments try and position development initiatives as adaptation programmes and spend large sums on consultants to establish additionality. The assessment of the Government of Nepal and ICIMOD is summarized as follows: 'What is urgently needed is a clear and fair definition of what additionality is in the context of mountain adaptation needs and in situations in which development activities are already present but need to be strengthened, given that climate change is exacerbating development problems' (GoN and ICIMOD 2010, pp. 6–7).

7.5 CONCLUSION

Mountains and coasts, which are at two ends of the terrestrial ecosystem, are most vulnerable to climate change impacts. While the coastal regions have received much attention in climate negotiations, mountains have not received due attention. The importance of mountain ecosystems and their services for society have been undervalued and as such mountain regions are the last frontiers for development due to their inaccessibility, marginality and fragility.

Nepal being a mountainous country provides the best case study to observe climate change and global warming occurring at higher elevations, and how it is affecting the vulnerability of the already poor mountain populations. The adverse impacts of climate change felt on agriculture and food security, water resources and hydropower, and biodiversity are critical for mountain ecosystems where warming rates are higher than in the low-lying lands. In addition, mountains are less developed areas that lack modern infrastructure and are characterized by limited livelihood options and limited freedom of choice. As the agriculture sector in the mountains is subsistence-oriented, rain-fed and in extremely small parcels of land, the rural populations are exposed to the vagaries of nature and are highly vulnerable to climate change impacts. One season of crop failure translates to severe hunger, especially in the already food-deficit mountain areas.

To mitigate these adverse impacts, the government has come up with an adaptation policy and strategies. Climate Change Policy (2011) and NAPA as an adaptation strategy are new instruments developed to comply with the global-level agreements under the aegis of the UNFCCC in order to be eligible for adaptation finance. In reality, the cost of adaptation is not estimated in a manner that reflects reality on the ground, leaving such policy and strategy at the mercy of donors, whether through the ODA or climate change finance.

While implementing NAPA, adaptation to climate change is seen in isolation by planners as an environmental issue and not linked within the realms of economy and development and vice versa. For example, involvement of the private sector does not feature in NAPA, and national-level poverty reduction programmes are not designed to be climate resilient. There is a dearth of capacity to dovetail adaptation planning into development planning and vice versa and this is going to be the most significant challenge in the future as the country tries to develop additional adaptation measures. Adaptation interventions also needs to take cognizance of long-term development goals that can contribute to the overarching goal of poverty reduction.

Though Nepal developed the Climate Change Policy and NAPA, adaptation finance has become a far-fetched dream for the most needy. The

GCF and REDD finance are good examples for LDCs where the funds are by and large illusive. Adaptation finance is often double counted with ODA. Like any other LDC, accessing climate change finance that is solely additional to ODA is the second biggest challenge as the qualification and compliance criteria are cumbersome and often not very straightforward and transparent.

NOTE

* The views expressed in this chapter are those of the author and not of the institution he is affiliated with.

REFERENCES

Akramov, K.T., B. Yu and S. Fan (2010), *Mountains, Global Food Prices, and Food Security in the Developing World*, Washington, DC: International Food Policy Research Institute (IFPRI).

Baidya, S.K., M.L. Shrestha and M.M. Sheikh (2008), 'Trends in daily climate extreme of temperature and precipitation in Nepal', *Journal of Hydrology and Meteorology*, **5** (1), 37–51.

Bolch, T., A. Kulkarni, A. Kääb et al. (2012), 'The state and fate of Himalayan glaciers', Science, **336** (6079), 310–14.

CPEIR (2011), *Nepal Climate Public Expenditure and Institutional Review*, Government of Nepal, National Planning Commission with support from UNDP/UNEP/CDDE in Kathmandu.

Dyhrenfurth, G.O. (1955), *To the Third Pole: The History of the High Himalaya*, London: Ex Libris, Werner Laurie.

Eriksson, M., X. Jianchu, A.B. Shrestha, R.A. Vaidya, S. Nepal and K. Sandstrom (2009), *The Changing Himalayas: Impacts of Climate Change on Water Resources and Livelihoods in the Greater Himalayas*, Kathmamdu: ICIMOD.

FAO (2012), *Statistics on Undernourishment: Prevalence of Undernourishment in Total Population (%)*, available at http://www.fao.org/economic/ess/ess-fs/fs-data/ess-fadata/en/ (accessed 16 August 2012).

Forrest, J.L., E. Wikramanayake, R. Shrestha et al. (2012), 'Conservation and climate change: assessing the vulnerability of snow leopard habitat to treeline shift in the Himalaya', *Biological Conservation*, **150**, 129–35.

Gallup, J.L., J.D. Sachs and A.D. Mellinger (1999), 'Geography and economic development', *International Regional Science Review*, **22**,179–232.

GIZ (2011), *Adaptation to Climate Change with a Focus on Rural Areas and India*, Duetsche Gesellschaft fur Internationale Zusammenarbeit, GmbH, India, Project on Climate Change Adaptation in Rural Areas of India, New Delhi.

GoN and ICIMOD (2010), Mountain Initiative Status Paper for UNFCCC and Rio+20 processes, Technical Expert Group of the Mountain Initiative, Government of Nepal and ICIMOD, Kathmandu.

Gosain, A.K., S. Rao and A.B. Shrestha (2011), 'Climate change impact assessment on water resources of Brahmaputra river basin', Paper presented at the Authors' Workshop for the Regional Report on Climate Change in the Hindu Kush–Himalayas: The State of Current Knowledge, ICIMOD, Kathmandu, 18–19 August.

Huddleston, B., E. Ataman, P. de Salvo et al. (2003), 'Towards a GIS-based analysis of mountain environments and populations', Environment and Natural Resources Working Paper 10, FAO, Rome.

ICIMOD (2012), 'Official submission to UNFCCC Secretariat by ICIMOD, Kathmandu (NWP member), proposal for a potential future area of work of the Nairobi Work Programme: Mountains and Climate Change', ICIMOD, Unpublished.

ISRC (2013), *District and VDC Profile of Nepal – 2013*, Kathmandu: Intensive Study and Research Centre.

Jodha, N.S. (1992), 'Mountain perspective and sustainability: a framework for development strategy', in N.S. Jodha, M. Banskota and Tej Partap (eds), *Sustainable Mountain Agriculture*, New Delhi and Oxford: IBH Publishing, pp. 41–82.

Kadota, T., K. Fujita, K. Seko, R.B. Kayastha and Y. Ageta (1997), 'Monitoring and prediction of shrinkage of a small glacier in the Nepal Himalaya', *Annuals of Glaciology*, **24**, 90–4.

LRMP (1986), *Land Utilization Report*, Kathmandu: Land Resources Mapping Project, Kenting Earth Sciences, His Majesty's Government of Nepal and Government of Canada.

Maplecroft (2012), *Climate Change Vulnerability Index (CCVI)*, available at http://maplecroft.com (accessed 1 September 2012).

Maskey, T.M. (1996), 'State of biodiversity in Nepal', Review Paper, in P. Shengie (ed.), *Banking on Biodiversity*, Reports on Regional Consultation on Biodiversity Assessment in the Hindu Kush and Himalayas, Kathmandu: ICIMOD.

Masters, W.A. and M.S. McMillan (2001), 'Climate and scale in economic growth', *Journal of Economic Growth*, **6** (3), 167–86.

McSweeney, C., M. New and G. Lizcano (2012), *Draft UNDP Climate Change Country Profiles: Nepal*, Oxford: School of Geography and the Environment, Oxford University.

MOE (2009), *Report of the High Level Task Force for Formulating 10 Yearly Hydropower Development of Nepal*, Kathmandu: Ministry of Energy, Government of Nepal.

MOEnv (2010), *National Adaptation Programme of Action (NAPA) to Climate Change*, Ministry of Environment, Kathmandu: Government of Nepal.

MOEST (2008), *Nepal Thematic Assessment Report: Climate Change*, National Capacity Needs Self Assessment for Global Environment Management Project (NCSA), Kathmandu: Ministry of Environment, Science and Technology (MOEST), Government of Nepal.

NBS (2002), *Nepal Biodiversity Strategy*, Kathmandu: Ministry of Forests and Soil Conservation.

NLSS (2004), *Nepal Living Standards Survey 2003/04*, Statistical Report Vol Two, Kathmandu: Central Bureau of Statistics, Government of Nepal.

OECD (2003), 'Development and climate change in Nepal: focus on water resources and hydropower', by S. Agrawala, V. Raksakulthai, M.V. Aalst, P. Larsen, J. Smith and J. Reynolds, Paris: OECD.

Qiu, J. (2008), 'The Third Pole', *Nature*, **454**, 393–6.

REDD-Forestry and Climate Change Cell (2011), *Role of Forest on Climate Change Adaptation*, Kathmandu: Ministry of Forests and Soil Conservation, Government of Nepal.

Schwank, O., A. Bruederle and N. North (2010), *Global Climate Financing Mechanisms and Mountain Systems*, Kathmandu: MOEnv and ICIMOD.

Shrestha, A. and R. Aryal (2011), 'Climate change in Nepal and its impact on Himalayan glaciers', *Regional Environmental Change*, **11** (1), 65–77.

Shrestha, A.B., C.P. Wake, P.A. Mayewski and J.E. Dibb (1999), 'Maximum temperature trends in the Himalaya and its vicinity: an analysis based on temperature records from Nepal for the period 1971–94', *Journal of Climate*, **12**, 2775–87.

Shrestha, A.B., C.P. Wake, J.E. Dibb and P.A. Mayewski (2000), 'Precipitation fluctuations in the Nepal Himalaya and its vicinity and relationship with some large scale climatological parameters', *International Journal of Climatology*, **20** (3), 317–27.

Singh, S.P., I. Bassignana-Khadka, B.S. Karky and E. Sharma (2011), *Climate Change in the Hindu Kush-Himalayas: The State of Current Knowledge*, Kathmandu: ICIMOD.

UNEP WCMC (2002), *Mountain Watch: Environmental Change and Sustainable Development in Mountains*, Nairobi: UNEP, available at http://www.unep-wcmc. org/mountains/mountainwatchreport/ (accessed 22 June 2010).

Viviroli, D., H.H. D'urr, B. Messerli, M. Meybeck and R. Weingartner (2007), 'Mountains of the world – water towers for humanity: typology, mapping and global significance', *Water Resource Research*, **43**, W07447, doi:10.1029/2006WR005653.

WECS (2005), *National Water Plan*, Kathmandu: Water and Energy Commission Secretariat (WECS), His Majesty's Government of Nepal.

WFPN (World Food Programme Nepal) (2009), *A Sub-regional Hunger Index for Nepal: Nepal Food Security Monitoring System*, Kathmandu: World Food Programme Nepal.

WHO (2012), http://www.who.int/nutrition/topics/3_foodconsumption/en/index. html (accessed 15 August 2012).

World Bank (2012), *World Development Indicators*, available at http://data. worldbank.org/ (accessed 1 September 2012).

8. Climate change adaptation in agriculture in Cambodia

Nyda Chhinh

Agriculture is a critical component of the Cambodian economy but agricultural production is frequently disrupted by flood and drought, which are likely to worsen due to climate change. This chapter provides an overview of the broad impact of climate change on Cambodia and, more specifically, the occurrence of flood and drought and their impact on paddy rice production in Cambodia. It then discusses current adaptation practices and policy interventions in the context of climate change by the Royal Government of Cambodia (RGC).

8.1 IMPACT OF CLIMATE CHANGE IN CAMBODIA

The Ministry of Environment (MoE) (2001) projects that climate change is going to alter Cambodia's rainfall in the wet season (May–November) as well as the dry season (December–April). It has used a Global Circulation Model to examine Cambodia's rainfall pattern and predicts that rainfall will vary up to 794 mm in the wet season. The finding of the MoE is echoed by McSweeney et al. (n.d.). They employed three climate change models to study Cambodia's rainfall patterns and find that climate change will result in increased rainfall between 0 to 14 per cent during the wet season and reduced rainfall during the dry season.

While the projection is made for the future climate of Cambodia, the cascading impact of climate change effects, especially the change in rainfall intensity and frequency, have already provided challenges for Cambodia's people and the country's development.

Paddy production has been impacted greatly by flood and drought. The distribution of average rainfall in Cambodia is between 1000 mm to 4000 mm per annum. Within the paddy rice production area in the flood plains, the rainfall is between 1000 mm and 1600 mm per annum. The terrain of Cambodia is often compared to a frying pan, with highlands and mountain ranges surrounding a flat country. This creates huge flood

163

plains around the Tonle Sap Lake and Mekong catchment. The Bassac River and the Mekong River systems potentially provide Cambodia with plentiful water for agriculture, especially paddy rice. However, the combination of a lack of physical infrastructure and the temporal and spatial shift in rainfall distribution in that area will spell disaster for agricultural production from both drought and flood.

The impact of climate change will intensify the hardship of people who are dependent on natural resources. Rainfall that has temporally shifted from the dry season to the wet season (McSweeney et al. n.d.) will unbalance the discharge of the Tonle Sap Lake and result in saltwater disturbance in the Mekong Delta area, which could disrupt agricultural production and cause damage to the social and physical infrastructure and ecosystems (Varis and Keskinen 2006; Vastila et al. 2010). The Tonle Sap Lake currently supports 2 million people directly and more than 6 million indirectly. When the shifting rainfall patterns are combined with large dam construction within the Mekong River basin, the Tonle Sap Lake's ecosystems are highly vulnerable (Benger 2007). It is believed that the Tonle Sap Lake is prone to climate change and the current capacity of the people who live in that area to adapt to environmental shock is weak (Nuorteva et al. 2010). Therefore, the intensity and frequency of the shocks can have disastrous consequences for the area and its inhabitants.

The impact of flood and drought in Cambodia has increased due to an increase in both intensity and frequency. The MoE (2001) confirms that between 1997 and 2001, rice production loss in Cambodia, one of the main contributors to economic growth, was associated with flood (about 70 per cent) and drought (about 20 per cent). A sharp decline in income from agriculture between 2001 and 2005 was also attributed to consecutive flood and drought events (Fitzgerald et al. 2007). Cambodian farmers have reported that flood and drought events have increased in both frequency and severity (Geres-Cambodia 2009; MoE 2005; MoE and UNDP 2011). It is evident that the effects of climate change are tangible and beyond the coping range of farmers. Policy development that will enhance the adaptive capacity of farmers to natural hazards, especially flood and drought that are worsening due to the changing climate, is urgently required.

8.2   AGRICULTURAL CONTEXT

The majority of the rice-growing provinces in Cambodia are located in a low-lying area called a 'lowland rice area' by agricultural experts. Situated within this region, the Mekong River plain region (which contains the Kampong Cham, Kandal, Phnom Penh, Prey Veng and Svay Rieng provinces) and the Tonle Sap plain region (which contains the Bantey Meanchey, Battambang, Kampong Chhnang, Kampong Thom, Pursat, Siemreap, Oddor Meanchey and Pailin provinces) account for 42 per cent and 42.5 per cent of Cambodia's total rice production area, respectively.

In Cambodia, rice is not only a staple food but also an economic good. Like many Asians, Cambodians eat rice at every meal. The World Food Programme (WFP) estimates that each Cambodian consumes around 150 kg of milled rice per year (RGC 2010b; WFP 2003). Cambodia must produce at least 3.5 million tons of paddy rice per year to feed its population of 14 million.[1] In 2009, Cambodia's total production of paddy rice was 7.59 million tons and export was expected to be 3 million tons after domestic consumption. However, the quantity of milled rice registered as exported was only 13,000 tons (RGC 2010b). Seeing rice as 'white gold' and a major growth opportunity for Cambodia, the RGC committed to turn the country into a 'rice-basket' (RGC 2010b). Rice is an important component of agriculture, which along with fisheries and forestry accounts for 28 per cent of gross domestic product (GDP) (NIS 2008). The share of services in the GDP is 38.3 per cent, while that of industry is 28.6 per cent. Unfortunately, rice production in Cambodia has faced many challenges.

The challenges in promoting agricultural productivity, mainly rice production, for farmers can be grouped into four categories: low capacity of farmers, high input cost, unstable output price and inadequate agricultural development.

8.2.1   Low Capacity of Farmers

Low productivity may be due to the low capacity of farmers to adapt to climate change. Cambodia's rice yield is comparatively low at around 2 tons per hectare. Vietnam and Thailand, in comparison, each produce approximately 3 tons per hectare (Yu and Fan 2011). As pointed out by Yusuf and Francisco (2010), Cambodia is vulnerable to climate change due to low adaptive capacity when exposed to natural hazards such as drought and flood. This low capacity of farmers can be due to farming techniques, inadequate investment and environmental reasons, such as rain-fed cultivation and poor soil.

In terms of farming techniques, farmers are still using traditional cultivation practices. A study by the MoE (2005) could not identify any measures taken by farmers to adapt to flooding other than traditional practices such as planting crops as usual with their traditional variety and moving to higher ground during flooding. Due to the fact that the majority of farmers (47 per cent) own less than 1 hectare of agricultural land (NIS 2009), employing modern farming techniques and using good-quality seeds would ensure that enough rice is produced to support household needs (Tsubo et al. 2009). However, there is slow adoption among farmers of modern farming techniques such as Systematic Rice Intensification (SRI) (MAFF 2011; Sovacool et al. 2012). According to Yu and Fan (2011), farmers who are trained in modern farming techniques yield about 16 per cent more than those who are not. Cambodia's Center for Study and Development in Agriculture (CEDAC) claims that improved knowledge about water and weed management, combined with planting different varieties of rice and using natural fertilizers, which farmers can make by composting, would double rice yield (CEDAC 2011). However, this approach is not frequently adopted in Cambodia (Deichert and Yang 2001). The number of farmers who did adopt it increased from 28 in 2000 to about 60,000 in 2006 (out of a total of 1.8 million farming households) (Yang 2012). This poor rate of take-up may be due to the increased labour intensity required for the new techniques. For example, water management for the SRI technique needs more monitoring and training than traditional methods.

Additionally, levels of investment by Cambodian farmers are relatively low. When cultivating rice in soil of poor quality, farmers must use more fertilizer (with appropriate application) to increase yield (Men 2007; World Bank 2006), however, studies show that Cambodian farmers fail to do this (Yu and Fan 2011). White et al. (1997) estimate that without fertilizer, soil in Cambodia will yield at maximum between 900 kg and 2600 kg per hectare.

The productivity of paddy rice in Cambodia is highly sensitive to changing rainfall due to the lack of irrigation systems and dikes to protect crops from drought and flood. About 30 per cent of the wet season rice is supplementary irrigated (NIS 2008). Physical damage to rice production caused by flood and drought is reported every year.

8.2.2 High Input Cost

Cost of fertilizer, pesticide and fuel, along with land rent, are the major input costs for paddy rice. Nam and Theng (2009) argue that there should be more low-rate loans available for farmers to enable them to invest in their production, stating that each 1 per cent increase in correctly used fertilizer leads to a 0.22 per cent increase in yield for dry season rice and 0.27 per cent for wet season rice. Touch and De Korte (2007) note that pests and disease have contributed to rice yield reduction dramatically and that, at the same time, if farmers use pesticide they can experience problems such as health damage and disability. The cost of pesticides may not be as high as the damage to health caused by pesticide use.

The cost of fuel is also a burden for Cambodian farmers. For example, during drought, farmers need to pump water into their paddy field. When this occurs, some farmers borrow cash at a high interest rate to buy diesel fuel or to hire pumps to irrigate their drying paddy field (McAndrew 1998).

The high rural–urban migration rates have created an agricultural labour shortage and require farmers to change from employing labour to renting machinery, thereby pushing the cost of production higher. In rural areas, there has been an increase in landless farmers who either engage in agricultural wage-labour or rent other people's land to cultivate paddy rice (Kim et al. 2002). The land rent is high, which reduces the profitability of cultivating rice (Agrifood Consulting International 2005).

8.2.3 Unstable Output Price

Farmers are hesitant to invest to increase yield due to the unreliable market price of paddy rice. Many poor farmers face a lack of cash before harvesting. In order to survive, they borrow money for almost all of their needs from middlemen and then have to sell their production soon after the harvest to those middlemen. Agrifood Consulting International (2005) finds that farmers often sell their production at the farm gate at well below the market price. If middlemen are involved in the transaction, the farmer must sell their paddy to the middlemen at an even lower price. Therefore, knowing that their paddy rice price will not be good, they apply a 'safety first' principle, which means that they invest only enough to secure their annual needs and thus compromise yield.

8.2.4 Inadequate Agricultural Development

There is lack of implementation of agricultural development programmes that promote rice productivity in Cambodia. The agricultural development that the government focuses on includes increasing farmer capacity, intervention on farming input costs and ensuring stable agricultural output prices, which are discussed in detail in Section 8.4.

When farmers' capacity is low (owing to reasons such as poor understanding of techniques, low investment and fragile soil) and government intervention is low, input prices can be high and farmers have difficulty dealing with unstable output prices.

These factors are treated with high priority in the *National Development Plan*, the major constraints of which are described in Section 8.3. Some government reports on the constraints, especially from the MoE, have confirmed that flood and drought are going to be major threats and require urgent intervention to address climate change impact in Cambodia (MoE

2005). Some authors claim that while Cambodia has abundant water in its rivers, lakes and aquifers, the livelihoods and food security of its people are threatened by drought (Turner et al. 2009).

8.3   NATURAL DISASTERS

The *Hyogo Framework for Action 2005–2015* posits that 'disaster risk arises when hazards interact with physical, social, economic and environmental vulnerabilities' (WCDR 2005, p. 1). Natural disasters are closely associated with natural hazards such as drought and flood (Chapman 1994). In the International Disaster Database,[2] maintained by the Centre for Research on the Epidemiology of Disaster (CRED, also known as ED-DAT), the disaster criteria include 'at least one of the following: 10 or more people killed, 100 or more people are affected, declaration of a state emergency or call for international assistance' (Guha-Sapir et al. 2012, p. 7). According to EM-DAT (2012) records, since 1900 Cambodia has experienced flooding 12 times, drought five times and tropical cyclones three times. The provincial-level records, however, contain more information than the CRED data and indicate that these events have occurred much more frequently.

Based on provincial data, Cambodia experienced flooding 20 times between 1984 and 2011 and flood has caused damage to paddy rice every year from 2000 to 2011. Damage to paddy rice caused by flood and drought is reported in the agricultural statistics of the Ministry of Agriculture, Forestry and Fishery (MAFF), especially from 1994. The compiled data of damage by drought and flood from various sources from 1984 to 2011 is shown in Figure 8.1. In total, drought damaged 1.09 million hectares and flood caused twice that damage between 1984 and 2011. The figure is greater than the cultivated area in 2010, which was about 2.5 million hectares.

It should be noted that Cambodia was experiencing civil war from the 1960s until the 1980s. Political stability has improved since 1993, when there was a general election supported by the United Nations Transition Authority in Cambodia. The data collected in this chapter is based on media reports, for example, articles from local newspapers, such as *SKP*, and the Agricultural Statistics Archive at the National Library and may not fully reflect the experience of Cambodian farmers at the time. However, the researchers have used multiple sources to ensure the data is as accurate as possible.

Flood and drought cause damage to paddy rice areas every year and are the most extreme climate events that occur in Cambodia (MoE

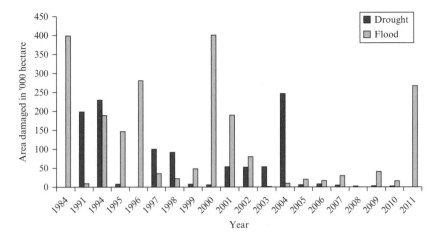

Figure 8.1   Flood and drought damage on paddy area in Cambodia, 1984–2011

2005). Natural disasters had a considerable impact on Cambodia during the 2000s. The most severe flooding occurred in 2000–01 and 2001–02, when floods hit the farmers hard in two consecutive years. The impact of drought was worst in 2004–05. As shown in Figure 8.1, between 2000 and 2005, Cambodia was impacted by both flood and drought. Flood caused extreme damage to paddy rice again in 2011.

8.3.1   Flood

Figure 8.2 shows the frequency of flooding events by province in Cambodia. The frequency is based on data on flood damage to paddy rice from various sources such as agricultural statistics. Based on the frequency of flooding, the provinces most seriously affected between 1988 and 2011 were Kandal, Kampong Thom, Battambang and Kratie. Prey Veng, Battambang and Takeo are located either along the Mekong River (the Plain region) or in the Tonle Sap region, and they were badly impacted by flood in 1996, 2000, 2001, 2002 and 2011. It is generally agreed that in 2000 (or 2000–01 based on crop productivity records), flooding was the worst in 70 years. In 2000, Prey Veng paddy rice area was severely damaged compared to other provinces. It is estimated that about 100 people were killed and there was US$170 million of agricultural losses due to flooding (Eng 2009).

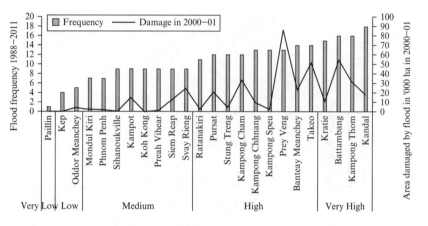

Note: We categorized frequency as high, medium or low based on the number of years that a province was impacted by flood from 1988 to 2011: Very Low (1–2), Low (3–6), Medium (7–10), High (11–14) and Very High (15–18).

Figure 8.2   Flood frequency by province 1988–2011 and area damaged by flood in 2000–01 in Cambodia

8.3.2   Drought

Figure 8.3 shows drought frequencies and associated damage to paddy rice area by province in Cambodia. Drought is recorded based on damage to areas of rice production caused by lack of water. There are three provinces that experienced drought very frequently, namely Kampong Speu, Takeo and Battambang. Prey Veng appears to be prone to both flood and drought, as the two worst disaster events in 2000–01 and 2004–05 badly damaged the paddy fields in that region. Based on annual drought information recorded by provinces in Cambodia between 1988 and 2011, the Kampong Speu, Takeo and Battambang provinces were considered the provinces most affected by drought.[3]

8.4   POLICY RESPONSE TO FLOOD AND DROUGHT

The aim of policy response to climate change, especially in relation to flood and drought, is to address the challenges faced by farmers. The policy is designed to create an environment that increases farmer capacity and ensures effective input price with a reliable market price for paddy rice. The capacity of farmers will be promoted by agricultural extension services and providing access to credit that enables them to invest more in

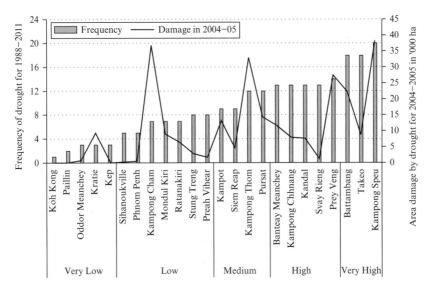

Figure 8.3   Drought frequency by province, 1988–2011 and area damaged by drought in 2004–05 in Cambodia

their paddy rice production. There should also be a crop insurance scheme to make sure that farmers have incentives to produce more rice.

Different government bodies approach flood and drought differently and with different action plans. For example, the main government authority responsible for natural disaster management is the National Committee for Disaster Management (NCDM), while the MoE is responsible for climate change resolution.

8.4.1   Institutional Arrangement

There are two ministries that directly address flood and drought issues in the agricultural sector (mainly with regard to paddy rice): the MAFF and the Ministry of Water Resource and Meteorology (MoWRAM). The MoE and the NCDM are also involved with flood and drought issues but with different approaches.

The MAFF and its line authorities (the provincial departments and district agriculture offices) primarily deal with agricultural services including training, supplying seeds and technical and marketing research. The MoWRAM and its line authorities are primarily concerned with water resources, such as building physical infrastructure and supplying large-scale pumping machines as needed. For example, when there is drought the Department of Agriculture provides new seedlings for farmers and the Department of Water Resource sends large machines to pump water into paddy fields. Regarding flood, the MoWRAM is responsible for pre-intervention, such as early warnings, while the MAFF is responsible for post-intervention, for example, assisting with restoring rice fields after flood damage. The MoWRAM also has a role in establishing Farmer Water User Committees (FWUCs) to manage water in their respective irrigation projects. A FWUC is not a government authority but a group formed by local residents around the area where each irrigation scheme is located.

The MoE is taking the lead in planning with regard to climate change issues, however, it is worth noting that the MoE has no line authority to deal with flood or drought. It has produced the *National Adaptation Programme of Action to Climate Change* (NAPA). The NAPA contains 20 high priority projects (the most crucial and urgent projects for immediate funding), including nine that respond to drought and five to flood. Within their existing roles and responsibilities, the MAFF and the MoWRAM are the implementation bodies for the NAPA.

The government agency that deals with human security and Disaster Risk Reduction (DRR) is the NCDM. Within this agency, the line authorities (Provincial, District and Commune Committee Disaster Management) are responsible for early warning, evacuation and DRR Strategic Plan and Actions.

Although Cambodia appears to have distinct government agencies to address flood and drought, there is usually cooperation among relevant agencies. Intergovernmental agencies are also formed to accomplish specific tasks. For example, to investigate and deal with the effects of climate change, the government issued Sub-Decree Number 35 in 2006 to establish the National Climate Change Committee, which has the prime minister as president, a minister from the MoE as chair and other high-ranking officials from relevant agencies as members. After the sub-decree was issued, working groups and task forces were formed as technical support to the National Committee.

8.4.2 Strategic Plans and Actions

The top-level development plan in Cambodia is the *National Strategic Development Plan* (NSDP). The NSDP's goal is to ensure that the country will achieve Cambodia's Millennium Development Goals.[4] The NSDP has 'good governance' as the central starting point, and it is hoped that achieving this aim will enhance development in the economy's major sectors, including agriculture. Discussion of the development involved in reaching the aim of good governance is beyond the scope of this chapter, however, it is central to the four components of the 'Rectangular Strategy', which includes (RGC 2010a, p. vi):

1. 'Enhancement of agriculture sector'.
2. 'Further rehabilitation and construction of physical infrastructure'.
3. 'Private sector development and employment generation'.
4. 'Capacity building and human resources development'.

Using these national overarching development guidelines as a basis, individual government authorities develop their own strategic development plans and actions. For example, the MAFF released the *Agricultural Sector Strategic Development Plan, 2006–2010*. The document states the comprehensive actions to safeguard against flood and drought planned for by the MAFF, including:

- water management and water supply
- land use planning and crop zoning
- soil fertility management and conservation
- research and development in crops and marketing.

While the MAFF is primarily responsible for agriculture and the MoWRAM for water, the two bodies have produced a joint agriculture development strategy called the *Strategy for Agriculture and Water 2006–2010*. The document contains two main strategies to address flood and drought issues: improving water resources, irrigation and land management; and improving agricultural and water research, education and extension. The MoWRAM has also drafted the *Climate Change Strategic Plan for Water Resources and Meteorology, 2013–2017*. This plan provides road maps to adapt and mitigate the effects of climate change in Cambodia.

There are a number of policies that, although not directly designed to deal with climate change adaptation, incorporate risk reduction strategies, such as the *Strategic National Action Plan* (SNAP) and the *Law on Water*. The SNAP, which is produced by the NCDM, addresses natural hazard risks including flood and drought. However, it mainly focuses on emergency arrangements and is not directly linked to climate change. *The Law on Water* of Cambodia deals with securing water for agriculture productivity and the welfare of people. Again, it has no direct association with climate change.

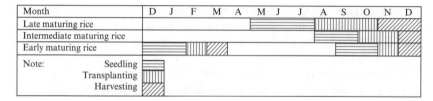

Figure 8.4   General paddy rice crop calendar in Cambodia

8.5   FLOOD AND DROUGHT ADAPTATION PRACTICES

Flood and drought can be examined in relation to the paddy rice crop production period. Typically, a one-year crop calendar covers an 11-month period that crosses two standard calendar years (Figure 8.4). For example, in 2013 the wet season rice grows from May to November 2013 and dry season rice from December 2013 to April 2014. Flood and drought occur only during the growth of wet season rice. Flooding usually occurs in September and drought may occur within any of the three rice-growing periods, during planting of seedlings, transplanting and harvesting.

8.5.1 Adaptation to Flood

The provinces that are in the two river catchments (the Mekong and Tonle Sap Rivers) are highly prone to the combination of flash flood and annual river flood.[5] From 1991 to 2011, half of all flash floods occurred within the second and third week of August and in September, which is during the period when rice plants are vulnerable. Flash flooding causes the rice plants to become weak and the annual flood intensifies and prolongs the flooding period. There are two rice ecosystems within the plain area that are known by local people as *sreleu* (which is grown at higher latitude) and *srekrom* (deep-water rice). The cultivation in the former area is rain-fed and the latter uses the hydrology of lakes and rivers (Javier 1997). Currently, farmers adjust the crop calendar to escape from drought and flood. For example, to avoid flash floods, farmers in rain-fed areas start their cultivation in May and harvest in August, but they may then be faced with drought. In deep-water rice cultivation, farmers cultivate before the arrival of flood waters as the plant will not grow much as the water rises. However, the rice yield is very low, with as little as 500 kg per hectare (Mak 2011).

Like drought, flash flooding can occur at any stage of the paddy growing period, but it is most destructive for plant yield if it occurs during the transplanting period (see Figure 8.4). At early growth stages, the plant needs to be submerged in the field for long periods and is vulnerable to the fast-flowing surface run-off that accompanies flooding.

A study conducted by the MoE (2001) suggests the options for agriculture to respond and adapt to climate change are the use of high-yield crop varieties, implementing early warning systems, improved irrigation, mapping flood and drought-prone localities, crop management, cultural practices and food diversification. These options were outlined in Cambodia's *Initial National Communication* to the United Nations Framework Convention on Climate Change (UNFCCC) in 2001 (MoE 2002). Early warning systems have been introduced at the local community level to monitor floods. However, they are mainly used for evacuation.

One promising area of adaptation to flooding is to use methods that result in high yields during flood and drought conditions. These are being tested by a number of scientists and have proved successful so far, but are still in the development and testing stage (Ikeda et al. 2008; MoE and UNDP 2011). Ikeda et al. (2008) indicate that using direct seedlings (seedlings that do not require tillage or transplanting) in different water conditions might lead to higher yields at a lower cost (as the farmer uses less labour to prepare the seedlings).

8.5.2 Adaptation to Drought

Based on current practices of paddy rice production (as shown in Figure 8.4), adaptation to drought must take into account three different types of drought, namely early-season drought, mid-season drought and late-season drought, which occur in June–July, August–September and October–November, respectively. There are two options widely discussed in the literature, namely changing seed varieties and building irrigation systems.

Changing seed varieties to adapt to drought
When farmers use late-maturing varieties of rice, it takes six months from planting seedlings to harvesting. For these farmers, the delayed onset of rainfall (May–July) is a drought event referred to as early-season drought. When such drought occurs, they may or may not change the seed varieties to intermediate-maturing varieties. The majority of farmers prefer late-maturing varieties, which are their traditional varieties of rice. By planting them the farmers take a risk. However, some farmers may wait until they have enough water in the rice field and use intermediate-maturing varieties. Such farmers must have two seed varieties to enjoy the flexibility of choosing between late- and intermediate-maturing varieties.

Cambodia normally experiences a period without rainfall for two weeks in early August. If this period stretches longer than two weeks, farmers who used late- and intermediate-maturing varieties experience drought that is called mid-season drought. If their paddy is damaged, the only option remaining for farmers is to use early-maturing varieties, but they are preferred least by farmers. Finally, there is late-season drought, which occurs when there is early cessation of rainfall (November and December) and affects all varieties of rice.

Water management
Drought can affect farmers regardless of which rice seed varieties they cultivate. Therefore, water management, such as building water reservoirs, must be considered. Farmers with access to water from irrigation schemes perform better than those lacking access, irrespective of whether there is drought. This is the most desirable option to mitigate the adverse effects of drought.

The RGC is promoting drought-resistant varieties of rice in addition to rehabilitating and constructing physical infrastructure, meaning that farmers can select which variety of rice to plant depending on the rainfall in a particular year.[6] Tsubo et al. (2009) find that when seeds are transplanted in June, farmers should select intermediate-maturity varieties so that the plant can avoid drought during the growing period and if sown in August, farmers should use early-maturing varieties.

In the case of rain-fed rice production, good rainfall during May and June induces farmers to practise their normal paddy rice cultivation. However, if this is followed by a dry spell of longer than two weeks in August, rice production is greatly impacted. As a first response to drought, farmers pump water from nearby sources such as ponds or tube wells. Without access to water, rice plants gradually wither and die and farmers must replant. However, poor, small landholder farmers usually have no seed remaining after the first planting. Currently, if this type of drought occurs and there is water nearby, government agencies take the steps necessary to save the rice plants by supplying small- and large-scale water pumps so that water can be transported to the rice fields. With the remaining destroyed rice fields, farmers can either wait for the appropriate conditions to plant early maturity rice varieties or leave rice fields uncultivated.

The most severe drought condition occurs when rainfall ceases during the grain-filling period (October to November) or the harvesting period. In this situation, the only possible plant-saving strategy is pumping water from water sources near the rice fields, but this becomes difficult or impossible when the drought is long and the nearby water bodies are unable to supply enough water.

MAFF (2011) and Sovacool et al. (2012) state that adaptation to flood and drought should come in three forms: changing seed varieties to those resistant to drought and flood; showing farmers how to use new farming methods, including integrated farming systems and the system of rice intensification (SRI); and providing training on how to build institutional, technical and capital resilience for climate-proof irrigation. Institutional resilience includes building the capacity of villagers to cope with drought and flood, and local authorities participating in the implementation of NAPA projects.

8.6 ADAPTATION DISCUSSION

There are at least two main challenges in agricultural production in Cambodia. First, natural hazards are more intense and occur more often. The MoE (2001) and IPCC (2007) agree that current trends will continue and climate change will affect rainfall distribution both spatially and temporally. The negative impacts of climate change will continue to occur. Second, agricultural development in Cambodia is relatively slow. Historically, Slocomb (2010) argues that Cambodia has failed to develop its agricultural sector. Overall, while the government and farmers are trying to address the challenges, flood and drought are still occurring and affecting agricultural development.

We believe that securing water for rice production is the most necessary and vital action for Cambodia to take to address the effects of climate change, followed by the other options listed in Section 8.5. While the crop varieties and crop calendar approach can be combined, without water no option is effective. At the same time, farming techniques such as SRI that can cope with paddy soil fertility and climatic conditions must be provided by scientific research. According to MAFF (2006), the constraints on rice production in Cambodia include an inadequate irrigation infrastructure, lack of agricultural inputs to improve soil fertility, limited access to education, training, credit and research, among many other enabling factors.

Despite existing work in agricultural development, paddy rice production has been affected by adverse climatic conditions. Perhaps it is time to revisit existing work and practices and examine how they could be altered to deal with the challenges of climate change. Given the openness of terrain in the flood plains of the two river systems, investment in infrastructure, such as dikes, may be costly. However, the government is implementing some initiatives to address this issue, as indicated in the NAPA.

It has been suggested that adaptation to climate change is no different from adapting to current climate hazards and that increasing resilience to climate hazards and reducing poverty are the key to increasing climate resilience in developing countries (Johnston et al. 2010). For Cambodia, adapting to floods means using flood-tolerant varieties of rice and, where possible, building dikes to prevent flooding. However, although these practices have been implemented, flood still damages rice production, as it did in 2011 when it destroyed crops on more than 10 per cent of Cambodia's total paddy area.

In short, many studies indicate that changing crop varieties and water management are the most common adaptive strategies being practised by households in Cambodia, in that early-maturing rice varieties are being used and rice plants are now established after floods have receded (MoE and UNDP 2011; Ouk et al. 2007; Ros et al. 2011). Since holding water in the rice fields is very critical for paddy rice, a common practice is for farms to raise their farm dike to hold water until the plants are ready to be harvested (ADPC 2005). However, an empirical study by Geres-Cambodia (2009) finds that changing crop calendars could also cause crop damage due to lack of rainfall towards the end of the rainy season and even households who use flood-resistant rice varieties are affected by drought.

8.7 CONCLUSION

Evidence of climate change effects in Cambodia is real and climate change models have predicted that effects will continue to worsen. In this chapter, we have demonstrated that agricultural production has been impacted by flood and drought. Between 1984 and 2011, on a national scale flood and drought were considered Cambodia's most severe natural disasters. The flooding in 2000 was the most severe in four decades and was repeated with similar intensity in 2001. In 2002, the country experienced both flood and drought.

The NAPA is a national adaptation action plan with minimal geographical coverage, which clearly defines Cambodia's planned adaptation activities, along with the specific location and budget estimate of each project proposed. In terms of institutional arrangements, the executive authority for managing water in Cambodia, appointed by the RGC, is the MoWRAM, which works in conjunction with a number of organizations from the government, civil society and the private sector. Where there are irrigation schemes at the community level, FWUGs are formed with technical support from the MoWRAM to ensure that water is used productively, wisely and in an equitable manner.

Adaptation activities include but are not limited to building dike and irrigation systems, which are recommended by almost every study on Cambodia's climate adaptability. Other recommendations advocated for adaptation include stronger agricultural extension services and changing inputs and crop varieties.

Structural adaptation measures proposed and implemented in Cambodia are focused on water management strategies to protect the country against drought and flood induced by climate change. These measures involve changing from rain-fed to irrigated agricultural practices. Although the NAPA is designed to be an adaptation action plan driven by climate change, it is effectively the same as other adaptation plans proposed in the literature about Cambodia related to adapting to drought. Research and development is needed to develop adaptation measures that can respond to climate variability, extremes and changes in long-term conditions affected by climate change.

In conclusion, the process of adapting to climate change in Cambodia is currently in an early stage. Agricultural production is still highly sensitive to changing rainfall and shifts in the duration and occurrence of rainfall have resulted in damage to production. This is closely associated with the low adaptive capacity of farmers due to their environmental, social and economic status. While experiencing the effects of climate change, farmers are also faced with high costs of inputs and an unstable price of output. This overall adaptive weakness requires improvement in government intervention, which is still at an early stage in terms of geographic coverage and practical options. From a government perspective, more should be done to ensure that farmers are able to cope with drought and flood including agricultural extension services, providing credit at low interest to farmers, enlarging the coverage of irrigation schemes, minimizing the cost of inputs and optimizing the outputs of farmers. If these conditions are met, farmers will have the environmental, social and economic capacity to cope with flood and drought.

ACKNOWLEDGEMENTS

This work was carried out with the aid of a grant from the International Development Research Centre (IDRC) in Ottawa, Canada and the Economy and Environmental Program for Southeast Asia (EEPSEA). The following people contributed to gathering data for this project: Cheb Hoeurn, Kong Sopheak, Sen Rineth, Noan Sereyroth and Cheang Sokhy. The comments and editing from Professor Sushil Vachani of Boston University are highly appreciated.

NOTES

1. Cambodia's population in 2010 was 14 million (NIS 2008).
2. EM-DAT is an emergency event database and can be accessed online at http://www.emdat.be/database.
3. The annual drought recorded is based on a number of sources including national agricultural statistics, newspaper reports, Government Disaster Declarations, National Disaster Committee Management, Asian Disaster Preparedness Centre and the Natural Disaster Database.
4. More information on the United Nations materials is available at http://www.un.org/millenniumgoals/.
5. 'Flash flood' refers to flooding within a short period caused by torrential rain in mountainous and highland areas that floods the plains. 'Annual river flood' is a seasonal flood caused by overflow of water in the river, especially the Tonle Sap and Mekong River.
6. Ten rice varieties promoted by the RGC from 2011 are: (1) Sen Pidao, (2) Chul'sa, (3) IR66 (these are early maturity), (4) Phka Rumdoul, (5) Khka Romeat, (6) Phha Romdeng, (7) Phka Chan Sen Sar (these are intermediate maturity), (8) Riang Chey, (9) CAR4 and (10) CAR6 (these are late maturity).

REFERENCES

ADPC (2005), *Vulnerability Assessment of Climate Risks in Strung Treng Province*, Phnom Penh: Asian Disaster Preparedness Center.

Agrifood Consulting International (2005), *Final Report for the Cambodia Agrarian Structure Study*, Phnom Penh: Agrifood Consulting International.

Benger, S. (2007), 'Remote sensing of ecological responses to changes in the hydrological cycles of the Tonle Sap, Cambodia', Paper presented at the Geoscience and Remote Sensing Symposium (IGARSS), IEEE International, Barcelona.

CEDAC (2011), *Climate Change Perception and Adaptation Practices by Rice-growing Communities in Cambodia*, Phnom Penh: Centre d'Etude et de Développement Agricole Cambodgien (CEDAC).

Chapman, D. (1994), *Natural Hazards*, Melbourne: Oxford University Press.

Deichert, G. and S.K. Yang (2001), 'Experiences with system of rice intensification in Cambodia', available at http://www.tropentag.de/2002/abstracts/datashows/110.pdf (accessed 20 October 2010).

EM-DAT (2012), *Cambodia Country Profile: Natural Disaster*, available at http://www.emdat.be/result-country-profile (accessed 12 December 2012).

Eng, R. (2009), 'Cambodia: mainstreaming flood and drought risk mitigation in East Mekong Delta', available at http://www.unescap.org/idd/events/2009_EGM-DRR/Cambodia-Eng-Rinbo-Flood-Drought-Risk-Mitigation-East-Mekong-Delta.pdf (accessed 7 December 2012).

Fitzgerald, I., S. So, S. Chan, S. Kem and S. Tout (2007), *Moving Out of Poverty? Trends in Community Well-being and Household Mobility in Nine Cambodia Villages*, Phnom Penh: Cambodia Development Resource Institute (CDRI).

Geres-Cambodia (2009), *Public Perceptions of Climate Change in Cambodia*, Phnom Penh: DanChurch Aid and Christian Aid.

Guha-Sapir, D., F. Vos, R. Below and S. Ponsrre (2012), *Annual Disaster Statistical Review 2011: The Number and Trends*, Brussels: CRED.

Ikeda, H., A. Kamoshita, J. Yamagishi, M. Ouk and B. Lor (2008), 'Assessment of management of direct seeded rice production under different water conditions in Cambodia', *Paddy and Water Environment*, **6** (1), 91–103.

IPCC (2007), *Climate Change 2007: Synthesis Report*, available at http://www.ipcc. ch/pdf/assessment-report/ar4/syr/ar4_syr.pdf (accessed 15 May 2012).

Javier, E.L. (1997), 'Rice ecosystems and varieties', in H.J. Nesbitt (ed.), *Rice Production in Cambodia*, Manila: IRRI, pp. 31–81.

Johnston, R., G. Lacombe, C.T. Hoanh et al. (2010), *Climate Change, Water and Agriculture in the Greater Mekong Subregion*, Colombo: International Water Management Institute.

Kim, S., S. Chan and A. Sarthi (2002), *Land, Rural Livelihoods and Food Security in Cambodia: A Perspective from Field Reconnaissance*, Phnom Penh: Development Resource Institute.

MAFF (2006), *Agricultural Sector Strategy Development Plan 2006–2010*, Phnom Penh: Ministry of Agriculture, Forestry and Fishery.

MAFF (2011), *Promoting Climate Resilient Water Management and Agricultural Practices in Rural Cambodia (NAPA Follow Up)*, Phnom Penh: Ministry of Agriculture, Forestry and Fishery.

Mak, S. (2011), 'Political geographies of the Tonle SAP: power, space and resources', PhD Thesis, National University of Singapore, Singapore.

McAndrew, J. (1998), *Interdependence in Household Livelihood Strategies in Two Cambodian Villages*, Phnom Penh: Cambodia Development Resource Institute.

McSweeney, C., M. New and G. Lizcano (n.d.), 'UNDP climate change country profiles: Cambodia', available at http://country-profiles.geog.ox.ac.uk (accessed 9 July 2012).

Men, S. (ed.) (2007), *Rice Book in Cambodia*, Phnom Penh: Cambodia Agricultural Research and Development.

MoE (2001), *Vulnerability and Adaptation Assessment to Climate Change in Cambodia*, Phnom Penh: Minstry of the Environment.

MoE (2002), *Cambodia's Initial National Communication*, Phnom Penh: Ministry of the Environment.

MoE (2005), *Vulnerability and Adaptation to Climate Hazards and to Climate Change: A Survey of Rural Cambodia Households*, Phnom Penh: Ministry of the Environment.

MoE and UNDP (2011), *Cambodia Human Development Report: Building Resilience the Future for Rural Livelihoods in the Face of Climate Change*, Phnom Penh: Ministry of the Environment.

Nam, T. and V. Theng (2009), 'Managing through the crisis – strengthening key sectors for Cambodia's future growth, development and poverty reduction: agriculture and rural development', *Cambodia Development Review*, **13** (2), 16–18.

NIS (2008), *Statistical Year Book of Cambodia 2008*, Phnom Penh: Ministry of Planning.

NIS (2009), *Cambodia Socio-economic Survey 2009*, Phnom Penh: Ministry of Planning.

Nuorteva, P., M. Keskinen and O. Varis (2010), 'Water, livelihoods and climate change adaptation in the Tonle Sap Lake area, Cambodia: learning from the past to understand the future', *Journal of Water and Climate Change*, **1** (1), 87–101.

Ouk, M., J. Basnayake, M. Tsubo et al. (2007), 'Genotype-by-environment interactions for grain yield associated with water availability at flowering in rainfed lowland rice', *Field Crops Research*, **101** (2), 145–54.

RGC (2010a), *National Strategic Development Plan Update 2009–2013*, Phnom Penh: Ministry of Planning.
RGC (2010b), *The Promotion of Paddy Production and Rice Export*, Phnom Penh: Royal Government of Cambodia.
Ros, B., P. Nang and C. Chhim (2011), *Agriculture Development and Climate Change: The Case of Cambodia*, Phnom Penh: CDRI.
Slocomb, M. (2010), *An Economic History of Cambodia in the Twentieth Century*, Singapore: NUS Press.
Sovacool, B.K., A.L. D'Agostino, A. Rawlani and H. Meenawat (2012), 'Improving climate change adaptation in least developed Asia', *Environmental Science and Policy*, **21**, 112–25.
Touch, V. and E. De Korte (2007), 'The current situation of chemical pesticide use on crops in Cambodia: is there any driving force to halt this application?', Paper presented at the Utilisation of Diversity in Land Use Systems: Sustainable and Organic Approaches to Meet Human Needs Conference, Witzenhausen.
Tsubo, M., S. Fukai, J. Basnayake and M. Ouk (2009), 'Frequency of occurrence of various drought types and its impact on performance of photoperiod-sensitive and insensitive rice genotypes in rainfed lowland conditions in Cambodia', *Field Crops Research*, **113** (3), 287–96.
Turner, S., G. Pangare and R.J. Mather (eds) (2009), *Water Governance: A Situational Analysis of Cambodia, Lao PDR and Viet Nam*, Vol. 2, Gland, Switzerland: IUCN.
Varis, O. and M. Keskinen (2006), 'Policy analysis for the Tonle Sap Lake, Cambodia: a Bayesian network model approach', *Water Resources Development*, **22** (3), 417–31.
Vastila, K., M. Kummu, C. Sangmanee and S. Chinvanno (2010), 'Modelling climate change impacts on the flood pulse in the Lower Mekong floodplains', *Journal of Water and Climate Change*, **1** (1), 67–86.
WCDR (2005), 'Hyogo Framework for Action 2005–2015: building the resilience of nations and communities to disasters', Paper presented to the World Conference on Disaster Reduction, Hyogo, Japan, available at http://www.unisdr.org/wcdr (accessed 25 May 2012).
WFP (2003), *Mapping Vulnerability to National Disasters in Cambodia*, Phnom Penh: National Committee for Disaster Management.
White, P.F., T. Oberthur and P. Sovuthy (1997), 'Rice in Cambodia economy: past and present', in H.J. Nesbitt (ed.), *Rice Production in Cambodia*, Manila: International Rice Research Institute, pp. 23–4.
World Bank (2006), *Cambodia: Halving Poverty by 2015?*, Washington, DC: World Bank.
Yang, S.K. (2012), 'Development of system rice intensification in Cambodia', available at http://www.cedac.org.kh (accessed 12 December, 2012).
Yu, B. and S. Fan (2011), 'Rice production response in Cambodia', *Agricultural Economics*, **42** (3), 437–50.
Yusuf, A.A. and H.A. Francisco (2010), *Mapping Climate Change Vulnerability in Southeast Asia*, Singapore: Economy and Environment Program for Southeast Asia (EEPSEA).

9. Adapting Indian agriculture to climate change

Sushil Vachani and Jawed Usmani

India is among the countries expected to bear the worst impact of climate change owing to its proximity to the equator and the vulnerability of a large segment of its population that lives in poverty. Climate change poses many challenges, but one of the greatest is the threat to India's ability to feed its growing population of 1.2 billion.[1] Apart from the impact of higher temperature, agriculture will be affected by water scarcity. This chapter focuses on Indian agriculture, its dependence on water resources and strategies for adaptation to climate change.

9.1 CLIMATE CHANGE IN INDIA

Annual mean temperature in India rose 0.056°C per decade between 1901 and 2007, but much faster in recent years (0.2°C per decade) (MEF 2012).[2] Temperature is forecast to increase 3–5°C by the end of the twenty-first century.[3] Warming will hurt India in many areas, with the most important being agriculture and water resources, energy and health. We briefly discuss the impact on energy and health, and then present a more detailed review of agriculture and water.[4]

9.1.1 Impact on Energy

The generation of energy is a significant source of greenhouse gases in India, as in many countries. Between 1953 and 2010, India's primary energy supply grew 3.8 percent per year and the share of coal in energy supply rose from 26 percent to 38 percent (MEF 2012, p. 23). Coal is a highly polluting energy source and increasing reliance on it exacerbates the problem of greenhouse gas emissions. Warming creates a vicious cycle leading to higher energy demand, greater energy generation, more emission of greenhouse gases and further warming. As temperature rises, demand for air-conditioning increases. As incomes grow, more people will demand air-conditioning in summer and heating in winter, though the need for heating will be tempered by milder winters. The additional energy requirement will be layered on top of demand arising from economic growth.

9.1.2 Impact on Health

When India became independent in 1947, the life expectancy of its people was a mere 32 years. By 2009, it had doubled to 64.5 years.[5] The rise can be attributed to better nutrition and access to health care, which, though still inadequate, has improved over the years. However, adverse effects of climate change threaten the future health of the Indian population. The larger number of very hot days will increase heatstroke. Greater rainfall in shorter periods will increase floods, which bring diseases such as diarrhea.

Higher temperature will cause malaria to spread into new regions. Malaria is transmitted by mosquitoes when temperature is between 18 to 32°C and humidity exceeds 55 percent (MEF 2012). With warming, some of the cold northern states, which have a low incidence of malaria at present, will be more exposed to it.[6] Some of the states to the south will have longer periods with malaria, while deeper south the transmission period will shorten. The incidence of malaria in India has dropped significantly in recent years,[7] but if this trend is to be sustained the country will have to take active measures to counter the effect of climate change. As part of its response, the government has highlighted the importance of refined forecasts that provide finer-grained projections that distinguish more meaningfully between locations with different geographic characteristics and account for variation in the population's socio-economic conditions, which affect their ability to safeguard against disease.

Extreme climate events such as heat waves, flood and drought will also disrupt livelihoods and worsen poverty for people struggling at the bottom of the pyramid, especially in rural areas where institutional shortcomings create severe disadvantages (Vachani and Smith 2008). Their level of nutrition, which is inadequate even at normal times, can be compromised and seriously injure health.

9.2 AGRICULTURE

While agriculture's share in India's gross domestic product (GDP) has declined steadily over the years to 18 percent in 2010, it still constitutes a critical part of the economy, providing livelihood for 650 million people.[8]

9.2.1 Indian Agriculture since Independence[9]

In the last 75 years, Indian agriculture has gone through disaster, restructuring and development, innovation and growth, consolidation and, finally, stagnation. Without radical improvement in productivity, India will fail to meet its growing demand for food even in the absence of the adverse effects of climate change. The demand for rice and wheat, the two main food grains consumed in India, is expected to rise to 225 million tonnes by 2020, as compared to the estimated production of around 181 million tonnes in 2010–11.[10] With climate change, the shortfall could be worse.

Disaster: the Great Bengal famine
The low point for Indian agriculture was the Great Bengal famine of 1943, when 3 million people starved to death primarily because they could not afford food as a result of wartime inflation.[11] An important lesson of the Bengal famine is that when a significant part of the population is extremely poor, and critically depends on agriculture for its survival, its ability to withstand price and income shocks is low and the government must devote special attention to protect it. This is also true today when, despite India's rapidly growing GDP in recent years, a large share of its population lives in poverty.[12]

Restructuring and development (independence to mid 1960s)
After independence, India focused on building infrastructure for agriculture. It invested in irrigation and manufacturing agricultural inputs. It set up agricultural research universities and extension services to provide information and support to farmers. Between 1950 and the mid 1960s, agriculture grew at around 2.4 percent per year,[13] but this was slower than population growth, which rose over 3 percent as the young country began providing its citizens with rudimentary health care (Swaminathan 2007). India was forced to import food from the USA.

Innovation and growth (mid 1960s to mid 1980s)
Having installed basic infrastructure, Indian agriculture could make leaps with innovation, such as high-yield wheat and rice, which, when combined with fertilizer and irrigation, yielded two to three times the grain as older varieties. Aggregate food production was also stimulated by institutional incentives. The National Bank for Agriculture and Rural Development was created to assist farmers with credit and coordinate rural development. The government set up mechanisms to help farmers market their produce and provided assured procurement prices. These measures combined to boost agricultural output in what came to be known as the Green Revolution.

During the early 1980s, agricultural growth exceeded population and GDP growth for the first time.

Consolidation (mid 1980s to 2000)
The last 15 years of the twentieth century are best characterized as a period of consolidation.[14] By 2004, India was the world's leading producer of tea, milk and fruit. It also had the second highest output of wheat, vegetables, fish and sugar. Even though its population grew to three times the size in 1951, its per capita production of grain rose 43 percent, milk output grew 68 percent and egg production jumped 500 percent.[15] While the Green Revolution and infrastructure investments since independence made India largely self-sufficient in food by the end of the twentieth century, as poverty remained high, not everyone got the necessary nutrition.

Stagnation (2000–10)
Despite the investments in infrastructure and institutions, and the remarkable productivity gains, agricultural growth has declined considerably in the past decade.[16] Food security is threatened by slower growth as well as continuing reliance on the monsoon, which causes production to swing wildly from year to year[17] (Figure 9.1).

Responding to the loss of momentum in agriculture, the Indian government announced the National Agricultural Policy in 2000 with the ambitious objective of raising growth to 4 percent per year. However, India

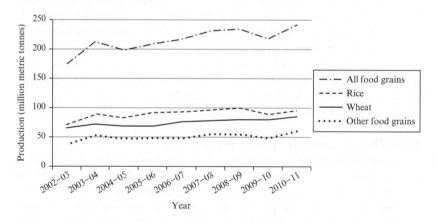

Source: Developed with data from table 4.1 (b), 'Target and achievement of production of major crops during Tenth (Xth) Five Year Plan (2002–03 to 2006–07) and 2007–08 to 2010–11', Ministry of Agriculture (2012a). The numbers for 2010–11 are estimates.

Figure 9.1   Production of major food grains in India

could not achieve that objective since soil fertility has eroded, water tables have fallen and water quality has dropped (MEF 2004). As late as 2009, only 48 percent of cultivated land was irrigated, leaving farmers at the mercy of the monsoon. The pace of providing irrigation has slowed.[18]

Uncertain future
By 2020, India's demand for food grain will be 30 percent to 50 percent higher than in 2012 (MEF 2012). The country may not be able to feed its population with domestically grown food without significant productivity gains, especially since farmers will be tempted to switch to producing for energy if that pays more.[19] Import dependence will create uncertainty since global supply will struggle to meet demand from the rising world population and face serious disruption from extreme climate events.[20] It is imperative for India to adapt to climate change and increase agricultural yield.

9.2.2   Effect of Climate Change on Agriculture[21]

Climate change will have a worse effect on countries such as India that are close to the equator than those further away (Cline 2008). In latitudes far from the equator, a rise in temperature can benefit agricultural production by lengthening the growing season.[22] The principal effects of climate change that can impact Indian agriculture include increase in temperature and carbon dioxide, change in rainfall pattern, higher incidence of drought and rise in sea level (MEF 2012).

Increase in temperature and carbon dioxide
Higher temperature increases evaporation of moisture from soil and leaves, hurting agricultural output. It also reduces yield by causing plants to mature fast, resulting in inadequate 'grain-filling' (Padgham 2009). Some of these negative effects can be countered by carbon fertilization benefits as atmospheric carbon dioxide increases. Simulations of the impact of rising temperature on Indian wheat production suggest that a 1°C rise in temperature would reduce output by 6 million tonnes, while a 5°C rise would cut it by 27.5 million tonnes without adjusting for carbon fertilization benefits and assuming no adaptation (MEF 2012). The combined overall effect of rising temperatures and carbon fertilization is uncertain.

Change in rainfall pattern
Rainfall varies significantly across India, from a high of over 1150 cm per year in the northeast to around 13 cm in the western state of Rajasthan (Attri and Tyagi 2010; MEF 2012). Climate projections indicate that by the end of this century, annual rainfall in India will be higher by 10 to

12 percent on average (MEF 2012). The change will vary depending on location and season. It is expected to increase in the east while decreasing on the Deccan plateau and Rajasthan. Rainfall will decrease between December and February and hurt the winter crop. The rest of the year will get more rain. The Indian government notes an 'alarming rise' in the intensity of the maximum one-day rainfall in the last 30 years (MEF 2012). Intense rain over short periods can cause flood and destroy crops.

Higher incidence of drought
Given India's high reliance on rain-fed agriculture, it is especially vulnerable to drought. In 2002–03, GDP growth slowed to 3.8 percent (from an average of 5.5 percent in the five preceding years) because drought hurt agricultural output (MEF 2012). About 16 percent of India's land is prone to drought (MEF 2012). Some crops, such as potato, are especially at risk. In the absence of adaptation, potato production could fall by 2.6 percent by 2020 and 15.3 percent by 2050 (MEF 2012). The projected impact is geographically uneven. Farmers in the states of Haryana, Punjab, Uttar Pradesh and Rajasthan could raise output by 3 to 7 percent by selecting appropriate varieties and adjusting the planting cycle. On the other hand, farmers in Gujarat, Maharashtra, Karnataka and Orissa could suffer losses of 6 to 46 percent. It is important to fund research to develop drought-resistant varieties to stem potential losses and to provide adaptation assistance tailored to the specific needs of different regions (MEF 2012).

Rise in sea level
By 2100, the sea level at India's coasts will have risen 3.5 to 34.6 inches compared with the 1990 level (MEF 2012). The destructive impact of storm surges will get compounded and threaten coastal agriculture by contaminating soil and groundwater with salt, and damaging coastal homes.

Regional variation
There will be significant variation in the effect of climate change in different parts of India. In the southwest, irrigated rice productivity is expected to fall by 4 percent in most of the Western Ghats, but further to the south it is forecast to rise. The effect will also depend on whether agriculture is on irrigated or rain-fed land. Where cultivation is on rain-fed land, rice yield could fall as much as 10 percent (MEF 2012). As over half of India's crops are grown on rain-fed land, which provides lower income and is more vulnerable to the harmful effects of climate change, it is important that farmers living off rain-fed land be given strong support.

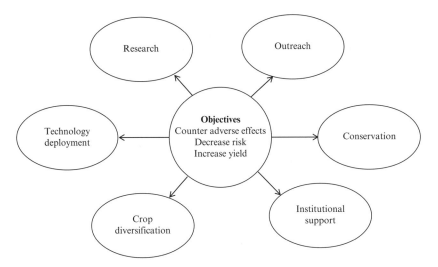

Figure 9.2   Objectives and strategies for adaptation of Indian agriculture to climate change

9.2.3   Adaptation to Climate Change in Agriculture[23]

India's strategies for adapting agriculture to climate change appear to have three broad objectives, countering adverse effects, decreasing risk and increasing yield. They can be classified into the six categories discussed below (Figure 9.2).

Technology deployment

The government is promoting agricultural techniques to reduce the adverse effects of climate change; for example, planting tomatoes on raised beds to protect them from floods or using crop varieties that can withstand drought. Losses in wheat production that normally stem from higher temperature can be limited by planting earlier than usual and using varieties suited to growing over a longer period. Use of low-cost methods for conserving soil moisture can decrease irrigation frequency and raise yield. These methods rely on use of inexpensive materials such as mulch derived from leaves and husk. Yield can also be stimulated with careful fertilizer usage. Genetic adaptation techniques have been used to develop varieties of coconut, eggplant and other crops that have high tolerance to stress such as high salinity, drought and extreme temperature (MEF 2012).

Research

India supports a network of publicly funded research institutions that focus on agriculture. There are many areas in which the need for further research has been identified. Varieties must be evaluated for their ability to survive higher temperature, drought and salinity. The government acknowledges the need to support research over an extended period as development of robust varieties can take several years (MEF 2012).

Crop diversification

Three crops, wheat, rice and maize, account for more than half of the food consumed in India (Government of India 2012). The use of high-yielding seeds has led farmers to concentrate on a smaller range of crops. Lower diversification of crops exposes farmers to higher risk from climate impact, disease and pests, and depletes soil nutrients more quickly. High-yielding wheat and rice consume more water and energy, both of which are scarce. A narrow range of crops also results in a less balanced diet for farming families (MEF 2012). Higher crop diversification can provide a measure of security even though it may reduce income in good years.

Conservation

The Indian government is promoting techniques to conserve water, improve efficiency of water usage and ensure output is not hurt by scarcity. Drip irrigation can help in reducing water usage in farming, especially for vegetables with shallow roots, such as tomato and onion. Greater use of mulch can help conserve soil moisture. In some areas, it may be possible to recycle sewage and industrial wastewater for agricultural use.

Conservation agriculture, which dispenses with conventional tillage for preparation of soil for sowing, and protects and enriches it with residue from previous harvests, has been identified as holding considerable promise for India (FAO 2012; Huggins and Reganold 2008; MEF 2012). The Food and Agriculture Organization (FAO 2012) describes it as a sustainable system based on the following principles:

1. 'Continuous minimum mechanical soil disturbance.'
2. 'Permanent organic soil cover.'
3. 'Diversification of crop species grown in sequences and/or associations'.

Over a period of several years, crop residue forms a protective cover for soil, creating 'a habitat for a number of organisms, from larger insects down to soil borne fungi and bacteria' that help create a favorable soil structure (FAO 2012, p. 2). Farmers must rotate crops and use integrated pest management techniques instead of conventional methods such as burning plant residue. (This provides the added benefit of limiting addition to the stock of greenhouse gases.) When farmers switch over to conservation agriculture, they have to be careful with use of pesticide and herbicide, so as not to disturb the organisms that have assumed the task of tillage. This requires technical assistance.

The benefits of conservation agriculture include earlier sowing (which can increase yield), lower use of fuel and fertilizer (which reduces cost) and fewer weeds. Furthermore, since the soil is not turned over there is lower release of carbon trapped in it and this helps reduce greenhouse gas emissions.

Outreach

Achievement of objectives underlying adaptation to climate change requires enormous programs for effective and sustained outreach to educate farmers and convince them to adapt practices. This is done partly by government-funded extension services supplemented by a large number of non-governmental organizations (NGOs), such as the Bharatiya Agro Industries Foundation (BAIF).[24] For example, NGOs are working with farming communities to demonstrate the benefits of optimal planting and persuade farmers to adopt efficient techniques. Agricultural NGOs have found that many farmers are misinformed about optimal spacing of plants and, in their anxiety to raise output, end up planting seeds too close, which reduces yield and raises costs. Some tend to apply more fertilizer and pesticide than necessary.

In some regions, there will be opportunities to benefit from climate change, for example, in Kerala changing conditions are likely to increase coconut yield (MEF 2012). By proactively planning for such changes, farmers can be assisted in maximizing gains by ensuring irrigation and soil enrichment. It is important to conduct extensive studies to project likely effects of climate change by region and district, and create comprehensive systems for disseminating information so that farmers can reduce risk and benefit from potential gains.

Institutional support

The government recognizes that farmers must be given stronger institutional support in a number of areas. Insurance products could help them safeguard against higher risks. Markets may not supply critical inputs such as special seeds and fertilizers most efficiently to all districts, requiring government oversight and intervention. Conservation can be encouraged by pricing water and energy to curb usage.

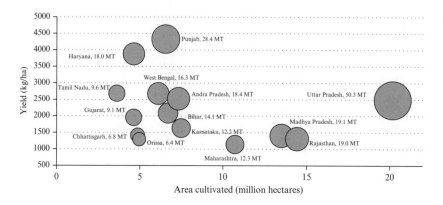

Note: The size of each circle corresponds to the output in million tonnes (MT), which is also noted next to the state's name. The data are the fourth advance estimates for 2011–12.

Source: Developed with data from table 4.5 (b), 'Area, production and yield of food grains during 2010–11 and 2011–12 in major producing states along with coverage under irrigation', Ministry of Agriculture (2012b), p. 63.

Figure 9.3   Agriculture: area under cultivation, output and yield by state, 2011–12

9.2.4   Bridging the Yield Gap

There are great differences in agricultural yield across Indian states, with Punjab and Haryana much ahead of others (Figure 9.3). Yield is a function of numerous factors including irrigation, choice of seeds, agricultural techniques, fertilizer use and farming techniques. Environmental conditions can prevent some regions from achieving yields possible in other places. Researchers determine potential yield for specific regions and identify region-specific yield gaps for different crops. The Indian government notes that potential yield for wheat and rice exceeds 6 tonnes per hectare, which is nearly three times the actual average yield of 2.06 tonnes per hectare (Ministry of Agriculture 2012b).

The overall agricultural output would increase substantially if states like Uttar Pradesh, which produces over 20 percent of India's food grain, could match the yield of Punjab, which was 74 percent higher than Uttar Pradesh's in 2011–12. If Uttar Pradesh's yield matched Punjab's, India's grain output would rise by 15 percent. Punjab's yield is only around 4.3 tonnes per hectare, so if the estimated potential yield of 6 tonnes per hectare could be achieved, the increase in output would be enormous. While it is impractical to expect such large increases, there does appear to be significant room for improving yield as indicated by studies based on modeling, experiments and actual maximum yield of selected farmers, which estimate potential yield over 6 tonnes per hectare for Uttar Pradesh, Punjab, Bihar and Assam, and between 5 and 6 tonnes per hectare for Andhra Pradesh, Karnataka, Orissa and Tamil Nadu (Lobell et al. 2009).

Strategies for improving yield are important irrespective of the anticipated effects of climate change as they would empower the population at the bottom of the pyramid, ensuring food security for the country and adding to its overall economic growth.

9.2.5 Implementation of Agricultural Adaptation Strategies

The plan for implementing India's adaptation strategies was put in place by the government in 2008 when it announced the *National Action Plan on Climate Change* (NAPCC), which presented eight missions that focused on different areas and promoted adaptation by incorporating sustainability principles in development planning (Table 9.1). The missions laid out broad objectives, such as encouraging market-based principles to boost energy efficiency.

The World Wildlife Fund-India viewed the NAPCC as 'a progressive and forward-looking document that ... show[ed] government's strong commitment to mainstream climate issues' and presented 'a positive message to public, industries and civil society about government's concern to address climate change through concerted action.'[25]

India has a number of policies and programs that aim to create a stronger and more sustainable agricultural sector, which will improve the ability of the enormous population dependent on agriculture to deal with the adverse effects of climate change. These aim to boost output, restore ecological balance, rejuvenate degraded soil and reduce inequity with measures such as improving irrigation and providing insurance (Table 9.2).

9.3 WATER

Water scarcity presents a great challenge for Indian agriculture, which consumes 83 percent of the country's water. Indians comprise 18 percent of the world's population, but survive on just 4 percent of its fresh water.[26] Between 1951 and 2007, water available per person fell by over two-thirds (MEF 2012). By 2050, the per capita amount of water will drop another 30 percent.

Most of India's water comes from rain and snow, which contributes about 4000 km^3 (billion cubic meters) of water each year; 80 percent of

Table 9.1   National missions under the National Action Plan on Climate Change (NAPCC)

No.	National mission	Objectives
1	National Solar Mission	Promoting ecologically sustainable growth while meeting energy security challenges
2	National Mission on Sustainable Habitat	Extension of Energy Conservation Building Code (optimization of energy demand) Urban planning/shift to public transport: long-term transport plans for small/medium cities Recycling of material and urban waste management: power from waste
3	National Mission on Green India	Double area under afforestation/ eco-restoration in the next ten years Increase greenhouse gas (GHG) removals by forests to 6.35% of India's annual GHG emissions by 2020 (increase of 1.5% over baseline) Enhance forests/ecosystems' resilience
4	National Mission for Sustaining the Himalayan Ecosystem	Strengthening institutional capacity Standardization of field and space observations Prediction/projection of future trends and assessment of possible impacts Governance for Sustaining Himalayan Ecosystem (G-SHE)
5	National Mission on Enhanced Energy Efficiency	Market-based approaches Cumulative avoidance of electricity capacity addition of 19,000 MW
6	National Water Mission	Conservation of water, minimizing wastage and ensuring its more equitable distribution both across and within states through integrated water resources development and management
7	National Mission for Sustainable Agriculture	Use of biotechnology Dryland (rain-fed) agriculture Risk management Access to information
8	National Mission on Strategic Knowledge on Climate Change	Network of institutions Promotion of climate science research Data sharing policy: from various arms of government Building human and institutional capacity: filling knowledge gaps in modeling and technology

Source: Selected data from Table 4.1, MEF (2012), pp.162–3. Reproduced with permission.

*Table 9.2   Policies/programs promoting sustainable development in the
             agriculture sector*

Policy/program	Feature
National Policy on Agriculture	Attain output growth rate in excess of 4% per annum based on efficient use of resources
Integrated Watershed Management Program	Restore ecological balance by harnessing, conserving and developing degraded natural resources such as soil, vegetative cover and water
National Watershed Development Project for Rain-fed Areas	Sustainable management of natural resources, enhancement of agricultural production, restoration of ecological balance in the degraded and fragile rain-fed ecosystems, reduction in regional disparity between irrigated and rain-fed areas and creation of sustained employment opportunities for the rural community including the landless
Rashtriya Krishi Vikas Yojana*	Assist states in the development and implementation of district-level agricultural plans (based on local agro-climatic conditions) and bring about quantifiable changes in the production and productivity of various components of agriculture and allied sectors
National Food Security Mission	Aims at increasing production of rice, wheat and pulses through area expansion and productivity enhancement in a sustainable manner; restoring soil fertility and productivity at the individual farm level
National Project on Organic Farming	Aims to promote production, promotion and market development of organic farming in the country
Micro Irrigation Scheme	Increase the area under efficient methods of irrigation like drip and sprinkler irrigation
Weather Based Crop Insurance Scheme	Aims to mitigate against the likelihood of financial loss on account of anticipated crop loss resulting from incidence of adverse conditions
National Horticulture Mission	To provide holistic growth of the horticulture sector through regionally differentiated strategies
National Project on Management of Soil Health and Fertility	Facilitate and promote Integrated Nutrient Management (INM) through judicious use of chemical fertilizers in conjunction with organic manures and bio-fertilizers

Note: * National Agricultural Development Program.

Source: Table 4.2, MEF (2012), p.167. Reproduced with permission.

Table 9.3   Water resources of major Indian rivers

River	Catchment area*	Potentially utilizable water resources		
		Surface	Ground	Total
	'000 km²	km³	km³	km³
Ganga	861	250.0	136.5	386.5
Godavri	313	76.3	33.5	109.8
Krishna	259	58.0	19.9	77.9
Mahanadi	141	50.0	13.6	63.6
Indus	321	46.0	14.3	60.3
Bhramaputra	194	24.3	25.7	48.0**
Narmada	99	34.5	9.4	43.9

Notes:

*   The numbers are for catchments in India. The Ganga, Indus and Bhramaputra also
     have catchment areas in other countries.
**   The numbers for surface and groundwater add up to 50. However, we have chosen to
     present the total as indicated in the source, which is 48.

Source: Selected data from Table 3.2, MEF (2012), pp. 104–5. Reproduced with
permission.

it is delivered by the southwest monsoon between June and October. An
estimated 500 km³ flows into India from neighboring countries, 210 km³
from Nepal, 195 km³ from China and 95 km³ from Bhutan (MWR 2008).
Much of the precipitation eventually flows through to the oceans. It is estimated that just 1,123 km³ (or 28 percent) can potentially be used. About 60
percent of this is surface water (mainly from rivers) and the rest is groundwater that is restored over time. Table 9.3 shows the rivers with the most
potential usable water. There is great variation in the share of rivers' total
water flow that can potentially be used; for example, the Brahmaputra has
about 10 percent greater water resources than the Ganga, but as indicated
in Table 9.3 the utilizable water from it is a fraction of that from the Ganga.

9.3.1   Impact of Climate Change on Water Resources

Simulations of overall demand and water availability suggest that while
total water availability exceeds demand, given variation over different parts
of the year and across regions, most parts of India are projected to experience 'crisis-like' conditions, which will worsen with climate change (MEF
2012, p. 113).[27] Vulnerability of water availability will vary with time and
across different river basins (Table 9.4). Unfortunately, the Ganga basin
downstream, which serves a large population, is highly vulnerable.

Table 9.4   Vulnerability of water availability in selected river basins

River basin	Year		
	2040	2070	2100
Ganga, upstream	Semi-vulnerable	Vulnerable	Vulnerable
Ganga, downstream	Highly vulnerable	Highly vulnerable	Highly vulnerable
Mahanadi	Semi-vulnerable	Vulnerable	Semi-invulnerable
Bhramaputra	Highly vulnerable	Highly vulnerable	Highly vulnerable

Source: Selected data from Table 3.3, MEF (2012), p. 111. Reproduced with permission.

Impact of melting glaciers on water resources

During spring and summer, snow melting in the Himalayas helps supplement flow in India's northern rivers. This is especially helpful in the summer months, preceding the monsoon, when there is little rain, and the levels of lakes, reservoirs and aquifers in the river basins are at their lowest.

Himalayan glaciers are receding as a result of rising temperature. For some years, this will increase flow into rivers, adding to the risk of flood when accompanied by torrential rain. Regions fed by Himalayan rivers have experienced destructive floods in recent years; for example, the devastation caused by floods in the Indian state of Uttarakhand in June 2013, in which over a thousand were killed (Peer 2013). Over time, however, the amount of water stored in glaciers will reduce significantly and their ability to supplement river flows in dry months will decrease. The seasonal snowfall in the Himalayas, which accumulates between October and February, and melts between February and September, will begin to melt earlier and constrain water availability in summer. The Indian government has identified the need to study differences in the region's microclimates, and their impact on different rivers, so that adaptation programs can be tailored accordingly (MEF 2012).

Impact of sea level rise on water resources

In addition to inundating significant coastal areas, causing property loss and displacing the population, sea level rise threatens agriculture by adding salt to soil and groundwater. As the coastal population uses larger amounts of fresh groundwater there is greater risk of drawing seawater into fresh water aquifers.[28]

9.3.2 Adaptation Strategies for Water

The main instrument for addressing the challenges of climate change for India's water resources is the National Water Mission, which lays out five goals (MWR 2011, p. iii):

(a) 'comprehensive water data base in public domain and assessment of impact of climate change on water resource'
(b) 'promotion of citizen and state action for water conservation, augmentation and preservation'
(c) 'focused attention to vulnerable areas including over-exploited areas'
(d) 'increasing water use efficiency by 20%'
(e) 'promotion of basin level integrated water resources management.'

It provides several action areas; for example, setting up networks for data gathering, collaborating with local elected officials, civil society and industry and timelines for achieving targets.

Adaptation in groundwater usage

Climate change will concentrate rainfall into fewer days giving it less time to percolate into the ground and recharge aquifers. So a larger share of rainfall will flow into rivers. At the same time, increase in population will enhance water demand, and more water will be lost to evaporation as temperature rises. In order to address future challenges effectively, the government will need to undertake extensive data collection and modeling of groundwater depletion and recharge patterns so that it can institute systems to manage water resources (MWR 2008). This is all the more necessary for areas where groundwater is overused at present.

The National Water Mission also highlights the necessity of clarifying policies and creating laws with regard to ownership and use of groundwater. It suggests that people who live on the land or own it should be allowed to draw groundwater for domestic consumption, and use limited amounts for irrigation without government regulation, except in exceptional circumstances such as drought. However, it asserts that government should regulate commercial use of groundwater and levy taxes on it (MWR 2008).

The Indian government set up the Central Groundwater Authority under the Environment (Protection) Act of 1986 to 'regulate and control development and management of groundwater resources.'[29] The Authority regulates groundwater withdrawal in industries, projects or geographic areas where it is assessed to be overexploited. In notified areas, only government agencies responsible for the supply of drinking water are allowed to build new tube wells. In order to encourage states to pass laws to control and manage groundwater development, the Ministry of Water Resources has suggested model legislation. So far six states and two union territories have enacted laws.

It is clear that in order to preempt severe water shortages, which could seriously compromise agricultural output, the Indian government will need to implement massive programs for water conservation, harvesting and storage so that more of the precipitation India receives can be retained for consumption in lean seasons, rather than flowing through to the sea, and what is available for consumption is used more efficiently.

9.4 CONCLUSION

While Indian agriculture has fared well since independence, it has begun to stagnate in the past decade with slowing growth in food production and smaller increase in irrigated land. Agricultural yields are still well below potential and there is significant variation across states. If yield could be raised in low-productivity states, such as Uttar Pradesh, the overall food supply would increase dramatically. This is easier said than done. While government officials are well informed about what needs to be done, it is not easy for them to implement programs that can change the perceptions and agricultural practices of a large number of poor farmers, some of whom live and work the way their forefathers did decades, if not centuries, ago. Mired in extreme poverty, they have little by way of resources to invest in improving yields, even if they were knowledgeable about the possibilities for enhancing output.

Climate change threatens food production, especially through water shortage. There will be substantial variation in impact across regions. However, there are also opportunities to benefit from climate change as higher carbon fertilization can raise the yield of some crops. Hence, there is a complex set of factors that are critical for the future of Indian agriculture and must be carefully understood and managed. As the government recognizes, research is needed in a number of areas, ranging from development of more robust crop varieties to managing water resources.

The government is addressing the challenges with a multi-pronged strategy to counter adverse effects, decrease risk and increase yield. In order to facilitate implementation of its strategy, it has set up eight missions that help focus attention on the need for adaptation and lay out specific actions and targets. The greatest challenge will be to reach out to the millions of farmers to convince them of the necessity to change agricultural practices to preempt the negative effects of climate change and take advantage of the opportunities. Success at adaptation will depend on how quickly and effectively the government can provide region-specific forecasts and work with civil society to convince farmers to adopt practices such as drip irrigation, conservation farming and use of stress-resistant seed varieties.

At a more macro level, the government must deploy resources to increase irrigation, both to improve yield and provide greater protection for poorer farmers. It will have to institute politically difficult policies such as pricing water and energy to encourage efficient use. Even if it does develop well-crafted policies, implementing them could be a challenge.

Overall, it appears that the effects of climate change could seriously compromise India's ability to feed its growing population, though the effects will vary across regions. The government has initiated a range of important programs to deal with the challenges. Many of the measures being contemplated are necessary to improve agricultural productivity and enhance food security for its poorest segments even if climate change were not a major issue.

NOTES

1. That was the population in 2011 as reported in Government of India (2011), p. ix.
2. Based on temperature data gathered at 121 locations around India by India's Meteorological Department.
3. The Indian government projected changes in climate for selected periods over this century with the British Hadley Center's high-resolution PRECIS modeling system.
4. Water will be scarce despite an anticipated increase in rainfall, which is forecast to increase 10 to 12 percent by the end of the twenty-first century (MEF 2012, p. 121).
5. Life expectancy at birth was 63 for males and 66 years for females in 2009 (World Bank 2011). The number for 1947 is provided in MEF (2012).
6. These include Jammu and Kashmir, Himachal Pradesh, Uttarakhand, Sikkim and Arunachal Pradesh.
7. The incidence of malaria in India dropped from 1.78 million reported cases in 2006 to 1.3 million in 2011, based on data obtained from the Global Health Observatory of the World Health Organization, available at http://apps.who.int/gho/data/view. country.10400 (accessed 4 April 2013).
8. Share of agriculture in GDP was reported to be 18 percent in 2010, based on data obtained from the World Development Indicators, World Bank, available at http://data bank.worldbank.org/data/views/variableSelection/selectvariables.aspx?source=world-development-indicators# (accessed 28 April 2013). About 46 percent of India's land is devoted to agriculture (MEF 2012, p. ii). For information on the number of people supported by agriculture, see MEF (2012), pp. 127 and 168.
9. This section draws on Swaminathan (2007).
10. The 2020 demand for rice was estimated at 122 million tonnes, and that for wheat at 103 million tonnes in MEF (2004), p. 83. In 2010–11, the production of rice was estimated to be 95.3 million tonnes and that of wheat, 85.9 million tonnes, as indicated in table 4.1(b), 'Target and achievement of production of major crops during Tenth (Xth) Five Year Plan (2002–03 to 2006–07) and 2007–08 to 2010–11', Ministry of Agriculture (2012a).
11. During that period India was a part of the British Empire. Amartya Sen (1981) explains that contrary to the government's assessment at the time that it was food short-

age that led to deaths, there was in fact more food in Bengal that year than in 1941. Unfortunately, there was very high inflation in the price of food as the government printed money to cover expenditure for the Second World War. While some people, especially those employed in government and living in Calcutta, saw their incomes rise and had access to a special allocation of food at controlled prices, which preserved their ability to purchase food, a large segment of the rural poor was simply unable to afford it. After remaining fairly steady since 1914, the price of rice soared 450 percent from December 1941 to May 1943, while the daily wage of unskilled agricultural laborers increased just 35 percent, slashing their purchasing power by 75 percent.

12. Starvation and malnutrition were not eliminated from India after it became independent. Thousands died of starvation in Bihar in 1966–67 and Maharashtra in 1973, though it is argued that the number of deaths was limited owing to government action (Rubin 2011).

13. Calculated from data on food production in table 4.5(a), 'All-India area, production and yield of food grains along with coverage under irrigation', Ministry of Agriculture (2012a), available at http://eands.dacnet.nic.in/latest_2006.htm (accessed 28 November 2012).

14. During this period, the focus shifted from wheat and rice to dairy, fruits, vegetables and oilseeds. Momentum for institutional development was sustained. In 1985, the National Wasteland Development Board was established to arrest land degradation and develop sustainable solutions for increasing land fertility. Information obtained from http://dolr.nic.in/iwdp1.htm (accessed 28 March 2011).

15. Based on data in MEF (2004), p. 83.

16. In the 1980s and 1990s, nine-year moving averages of growth were as high as 4 percent to 5 percent in some periods whereas in the last decade they have been 2 percent or less. Calculated from data provided in table 4.5(a), 'All-India area, production and yield of food grains', Ministry of Agriculture (2012a), available at http://eands.dacnet.nic.in/latest_2006.htm (accessed 28 November 2012).

17. In 16 of the last 60 years, production has changed by more than 10 percent compared to the previous year. In five of those years, it changed by over 20 percent in a year.

18. From 1998–99 to 2008–09 the percentage of irrigated land rose by 5.9 percent, which was much slower than during the previous decade when it increased by 8 percent. Calculated from data in table 4.5(a), 'All-India area, production and yield of food grains along with coverage under irrigation', Ministry of Agriculture (2012a), available at http://eands.dacnet.nic.in/latest_2006.htm (accessed 28 November 2012).

19. Based on data provided in Swaminathan (2007), between 2005 and 2007, the price of corn (which is used for production of ethanol in addition to serving as food) rose 66 percent, while the price of wheat rose only 34 percent.

20. By 2050, there are estimated to be 9.3 billion people on Earth compared with 7 billion today. Estimated from the US Census Bureau Database, available at http://www.census.gov/population/international/data/idb/region.php (accessed 3 December 2012). The world will have to increase agricultural output by 60 percent to feed its population and provide inputs for biofuels. Most of the increase must come from higher productivity, as the amount of additional land that can be used for agriculture is only about 5 percent (OECD/FAO 2012).

21. This section draws significantly from MEF (2012).

22. Regions that will benefit include northern parts of North America and China, the southern regions of South America and higher elevations of East Africa (Padgham 2009).

23. This section draws significantly from MEF (2012).

24. See http://www.baif.org.in.

25. 'WWF-India's reaction to the National Climate Change Action Plan', available at http://www.wwfindia.org/news_facts/infocus/index.cfm (accessed 1 December 2012).

26. MEF (2012), pp. 5 and 103 refers to data from the UN Population Database.

27. Modeling of the effects of climate change on water resources indicates that most river

basins (with the exception of the Cauvery, Bhramaputra and Pennar) are projected to receive more rain, but experience higher evaporation loss from soil and vegetation. The Indian government modeled the effects of climate change on water resources with the Soil and Water Assessment Tool (SWAT) based on data on land use, soil and terrain along with weather data generated by the Indian Institute of Tropical Meteorology, Pune using PRECIS simulations. SWAT is described as 'a river basin, or watershed, scale model' developed for the United States Department of Agriculture (USDA) Agricultural Research Service (ARS) to 'predict the impact of land management practices on water, sediment and agricultural chemical yields in large complex watersheds with varying soils, land use and management conditions over long periods of time' (Neitsch et al. 2011, p. 1).

28. For more on this topic, see figure 4, p. III/18, MWR (2008).
29. The material in this paragraph is based on http://cgwb.gov.in/gw_regulation.html (accessed 5 December 2012).

REFERENCES

Attri, S.D. and A. Tyagi (2010), *Climate Profile of India*, New Delhi: Indian Meteorological Department.

Cline, W.R. (2008), 'Global warming and agriculture', *Finance and Development*, 23–27 March, available at http://www.imf.org/external/pubs/ft/fandd/2008/03/pdf/cline.pdf (accessed 9 April 2013).

FAO (Food and Agriculture Organization) (2012), *What is Conservation Agriculture?*, available at http://www.fao.org/ag/ca/1a.html (accessed 20 November 2012).

Government of India (2011), *Census of India: Provisional Population Totals, Paper 1 of 2011*, New Delhi: Office of the Registrar General and Census Commissioner, available at http://www.censusindia.gov.in/2011-prov-results/data_files/india/paper_contentsetc.pdf (accessed 15 April 2013).

Government of India (2012), *Economic Survey*, available at http://indiabudget.nic.in/survey.asp (accessed 3 December 2012).

Huggins, D.R. and J.P. Reganold (2008), 'No-till: the quiet revolution', *Scientific American*, **299** (1), July, 70–7.

Lobell, D.B., K.G. Cassman and C.B. Field (2009), 'Crop yield gaps: their importance, magnitudes and causes', *Annual Review of Environment and Resources*, **34**, 179–204.

MEF (Ministry of Environment and Forests) (2004), *India's National Communication to the United Nations Framework Convention on Climate Change*, New Delhi: Ministry of Environment and Forests, Government of India.

MEF (Ministry of Environment and Forests) (2012), *India, Second National Communication to the United Nations Framework Convention on Climate Change*, New Delhi: Ministry of Environment and Forests, Government of India.

Ministry of Agriculture (2012a), *Agricultural Statistics at a Glance, 2011*, New Delhi: Ministry of Agriculture, available at http://eands.dacnet.nic.in/latest_2006.htm (accessed 1 December 2012).

Ministry of Agriculture (2012b), *Agricultural Statistics at a Glance, 2012*, New Delhi: Ministry of Agriculture, available at http://eands.dacnet.nic.in/Publication12–12–2012/Agriculture_at_a_Glance%202012/Pages38–84.pdf (accessed 2 April 2013).

MWR (Ministry of Water Resources) (2008), *Comprehensive Mission Document, Volume 2 (Draft)*, New Delhi: National Water Mission under National Action Plan on Climate Change, Ministry of Water Resources, available at http://wrmin. nic.in/writereaddata/linkimages/Mission_Doc_Vol22880755143.pdf (accessed 5 December 2012).

MWR (Ministry of Water Resources) (2011), *Comprehensive Mission Document, Volume 1*, New Delhi: National Water Mission under National Action Plan on Climate Change, Ministry of Water Resources, available at http://india.gov.in/ allimpfrms/alldocs/15658.pdf (accessed 5 December 2012).

Neitsch, S.L., J.G. Arnold, J.R. Kiniry and J.R. Williams (2011), *Soil and Water Assessment Tool Theoretical Documentation Version 2009*, Texas: Texas A&M University, available at http://twri.tamu.edu/reports/2011/tr406.pdf (accessed 4 December 2012).

OECD/FAO (2012), *OECD-FAO Agricultural Outlook 2012–2021*, Paris: OECD Publishing, available at http://dx.doi.org/10.1787/agr_outlook-2012-en (accessed 3 December 2012).

Padgham, J. (2009), *Agricultural Development under a Changing Climate: Opportunities and Challenges for Adaptation*, Washington, DC: World Bank.

Peer, B. (2013), 'Weather hampers rescue of flood victims in India', *New York Times*, 24 June, p. 10.

Rubin, O. (2011), *Democracy and Famine*, Abingdon, Oxon and New York: Routledge.

Sen, A. (1981), *Poverty and Famines: An Essay on Entitlement and Deprivation*, Oxford: Oxford University Press.

Swaminathan, M.S. (2007), 'The crisis of Indian agriculture', *Hindu*, available at http://www.hinduonnet.com/af/india60/stories/2007081550320900.htm (accessed 26 March 2011).

Vachani, S. and N.C. Smith (2008), 'Socially responsible distribution: strategies for reaching the bottom of the pyramid', *California Management Review*, **50** (2), 52–84.

World Bank (2011), *World Development Report, 2012*, Washington, DC: World Bank.

Index